Praise for

LOST IN MATH

"Hossenfelder ably mixes simplified explanation of the science with compelling portraits of the fascinating characters who study it."

—*Vanity Fair*

"In her new book, *Lost in Math*, Sabine Hossenfelder adroitly confronts this crisis head on.... The book is a wild, deep, thought-provoking read that would make any reasonable person in the field who's still capable of introspection doubt themselves." —*Forbes*

"*Lost in Math* is self-aware and dosed with acerbic wit, and it asks bold questions."

—*Nature*

"Sabine Hossenfelder's new book, *Lost in Math*, provides a well-informed take on the current situation in fundamental physical theory. The author is completely honest, utterly fearless, and often quite funny."

—*MAA Reviews*

"[Hossenfelder's] critical assessment of the field is appropriately timed."

—*Science*

"Eavesdrop on accessible and frank conversations in Hossenfelder's *Lost in Math*, which wrestles with big questions of quantum mechanics and beauty in a fun, fascinating way." —*Popular Science*

"Entertaining and engaging." —*Ars Technica*

"Hossenfelder's jaunt through the world of theoretical physics explicitly raises the question of whether the activities of thousands of physicists should actually count as 'science.' And if not, then what in tarnation are they doing?" —*Weekly Standard* (UK)

"Even educated readers will struggle to understand the elements of modern physics, but they will have no trouble enjoying this insightful, delightfully pugnacious polemic about its leading controversy."

—*Kirkus* (starred review)

"This layreader-friendly, amusing treatise gives an enlightening look at a growing issue within physics." —*Publishers Weekly*

"Emphasizing how much researchers have achieved in quantum mechanics while using math that is decidedly ugly, Hossenfelder urges her colleagues to start focusing on reality, not conceptual style. A provocative appeal for unattractive but fruitful science."

—*Booklist*

"Hossenfelder, a philosophically inclined physicist, presents the informed reader with a fascinating panorama of the current state of physics, replete with imaginative entities like wormholes, parallel universes, and bubbles associated with the baby universe whose existence cannot be established or falsified through standard experimental modes." —*CHOICE*

"Born too late to savor the heady era when the standard model of particle physics came together, Sabine Hossenfelder is impatient for new waves of discovery. Might the pace of insights be slowing because illusions of mathematical beauty have beguiled her fellow theorists? *Lost in Math* chronicles her quest—through interviews and conversations—to set her own course for exploration."

—Chris Quigg, distinguished scientist emeritus,
Fermi National Accelerator Laboratory

"*Lost in Math* is a delight. It is engaging, witty, and utterly profound. If you want to know why so many contemporary theoretical physicists choose to believe unbelievable things, this is a great place to start."

—Jim Baggott, author of *Farewell to Reality*

"Centered around insightful interviews with leading theorists, *Lost in Math* provides a well-informed take on the current state of fundamental physical theory from a physicist who is utterly fearless, completely honest, and quite funny."

—Peter Woit, mathematical physicist at Columbia University and author of *Not Even Wrong*

Also by Sabine Hossenfelder

Experimental Search for Quantum Gravity (editor)

LOST IN MATH

HOW BEAUTY LEADS PHYSICS ASTRAY

SABINE HOSSENFELDER

BASIC BOOKS
New York

Basic Books
Hachette Book Group
1290 Avenue of the Americas, New York, NY 10104
www.basicbooks.com

Printed in the United States of America

Originally published in hardcover and ebook by Basic Books in June 2018
First Trade Paperback Edition: June 2020

Published by Basic Books, an imprint of Perseus Books, LLC, a subsidiary of Hachette Book Group, Inc. The Basic Books name and logo is a trademark of the Hachette Book Group.

The Hachette Speakers Bureau provides a wide range of authors for speaking events. To find out more, go to www.hachettespeakersbureau.com or call (866) 376-6591.

The publisher is not responsible for websites (or their content) that are not owned by the publisher.

Print book interior design by Jouve

Library of Congress Cataloging-in-Publication Data
Names: Hossenfelder, Sabine, 1976- author.
Title: Lost in math: how beauty leads physics astray / Sabine Hossenfelder.
Description: First edition. | New York: Basic Books, [2018] | Includes bibliographical references and index.
Identifiers: LCCN 2017057165| ISBN 9780465094257 (hardcover) | ISBN 0465094252 (hardcover) | ISBN 9780465094264 (ebook) | ISBN 0465094260 (ebook)
Subjects: LCSH: Mathematical physics. | Cosmology. | Quantum theory.
Classification: LCC QC20.6 .H67 2018 | DDC 530.15—dc23
LC record available at https://lccn.loc.gov/2017057165

ISBNs: 978-0-465-09425-7 (hardcover); 978-1-5416-4676-6 (paperback); 978-0-465-09426-4 (ebook)

LSC-C

Printing 4, 2024

To Mom

Heaven holds a sense of wonder

And I wanted to believe

That I'd get caught up

When the rage in me subsides

—SARAH MCLACHLAN, "SILENCE"

CONTENTS

PREFACE

THEY WERE so sure, they bet billions on it. For decades physicists told us they knew where the next discoveries were waiting. They built accelerators, shot satellites into space, and planted detectors in underground mines. The world prepared to ramp up the physics envy. But where physicists expected a breakthrough, the ground wouldn't give. The experiments didn't reveal anything new.

What failed physicists wasn't their math; it was their choice of math. They believed that Mother Nature was elegant, simple, and kind about providing clues. They thought they could hear her whisper when they were talking to themselves. Now Nature spoke, and she said nothing, loud and clear.

Theoretical physics is the stereotypical math-heavy, hard-to-understand discipline. But for a book about math, this book contains very little math. Strip away equations and technical terms and physics becomes a quest for meaning—a quest that has taken an unexpected turn. Whatever laws of nature govern our universe, they're not what physicists thought they were. They're not what I thought they were.

Lost in Math is the story of how aesthetic judgment drives contemporary research. It is my own story, a reflection on the use of what I was taught. But it is also the story of many other physicists who struggle with the same tension: we believe the laws of nature are beautiful, but is not believing something a scientist must not do?

1
The Hidden Rules of Physics

In which I realize I don't understand physics anymore. I talk to friends and colleagues, see I'm not the only one confused, and set out to bring reason back to Earth.

The Conundrum of the Good Scientist

I invent new laws of nature; it's what I do for a living. I am one of some ten thousand researchers whose task is to improve our theories of particle physics. In the temple of knowledge, we are the ones digging in the basement, probing the foundations. We inspect the cracks, the suspicious shortcomings in existing theories, and when we find ourselves onto something, we call for experimentalists to unearth deeper layers. In the last century, this division of labor between theorists and experimentalists worked very well. But my generation has been stunningly unsuccessful.

After twenty years in theoretical physics, most people I know make a career by studying things nobody has seen. They have concocted mind-boggling new theories, like the idea that our universe is but one of infinitely many that together form a "multiverse." They have invented dozens of new particles, declared that we are projections of a higher-dimensional space, and that space is spawned by wormholes that tie together distant places.

These ideas are highly controversial and yet exceedingly popular, speculative yet intriguing, pretty yet useless. Most of them are so difficult to test, they are practically untestable. Others are untestable even theoretically. What they have in common is that they are backed up by theoreticians convinced that their math contains an element of truth about nature. Their theories, they believe, are too good to not be true.

The invention of new natural laws—theory development—is not taught in classes and not explained in textbooks. Some of it physicists learn studying the history of science, but most of it they pick up from older colleagues, friends and mentors, supervisors and reviewers. Handed from one generation to the next, much of it is experience, a hard-earned intuition for what works. When asked to judge the promise of a newly invented but untested theory, physicists draw upon the concepts of naturalness, simplicity or elegance, and beauty. These hidden rules are ubiquitous in the foundations of physics. They are invaluable. And in utter conflict with the scientific mandate of objectivity.

The hidden rules served us badly. Even though we proposed an abundance of new natural laws, they all remained unconfirmed. And while I witnessed my profession slip into crisis, I slipped into my own personal crisis. I'm not sure anymore that what we do here, in the foundations of physics, is science. And if not, then why am I wasting my time with it?

∞∞

I WENT INTO physics because I don't understand human behavior. I went into physics because math tells it how it is. I liked the cleanliness, the unambiguous machinery, the command math has over nature. Two decades later, what prevents me from understanding physics is that I still don't understand human behavior.

"We cannot give exact mathematical rules that define if a theory is attractive or not," says Gian Francesco Giudice. "However, it is surprising how the beauty and elegance of a theory are universally recognized by people from different cultures. When I tell you, 'Look, I have a new paper and my theory is beautiful,' I don't have to tell you the details of my theory; you will get why I'm excited. Right?"

I don't get it. That's why I am talking to him. Why should the laws of nature care what I find beautiful? Such a connection between me and the universe seems very mystical, very romantic, very not me.

But then Gian doesn't think that nature cares what *I* find beautiful, but what *he* finds beautiful.

"Most of the time it's a gut feeling," he says, "nothing that you can measure in mathematical terms: it is what one calls physical intuition. There is an important difference between how physicists and mathematicians see beauty. It's the right combination between explaining empirical facts and using fundamental principles that makes a physical theory successful and beautiful."

Gian is head of the theory division at CERN, the Conseil Européen pour la Recherche Nucléaire. CERN operates what is currently the largest particle collider, the Large Hadron Collider (LHC), humankind's closest look yet at the elementary building blocks of matter: a $6 billion, 16-mile underground ring to accelerate protons and smash them together at almost the speed of light.

The LHC is a compilation of extremes: supercooled magnets, ultrahigh vacuum, computer clusters that, during the experiments, record about three gigabytes of data—comparable to several thousand ebooks—per second. The LHC brought together thousands of scientists, decades of research, and billions of high-tech components for one purpose: find out what we're made of.

"Physics is a subtle game," Gian continues, "and discovering its rules requires not only rationality but also subjective judgment. For me it is this unreasonable aspect that makes physics fun and exciting."

I am calling from my apartment, cardboard boxes are stacked around me. My appointment in Stockholm has come to an end; it's time to move on and chase another research grant.

When I graduated, I thought this community would be a home, a family of like-minded inquirers seeking to understand nature. But

I have become increasingly alienated by colleagues who on the one hand preach the importance of unbiased empirical judgment and on the other hand use aesthetic criteria to defend their favorite theories.

"When you find a solution to a problem you have been working on, you get this internal excitement," says Gian. "It is the moment in which you suddenly start seeing the structure emerging behind your reasoning."

Gian's research has focused on developing new theories of particle physics that hold the prospect of solving problems in existing theories. He has pioneered a method to quantify how natural a theory is, a mathematical measure from which one can read off how much a theory relies on improbable coincidences.[1] The more natural a theory, the less coincidence it requires, and the more appealing it is.

"The sense of beauty of a physical theory must be something hardwired in our brain and not a social construct. It is something that touches some internal chord," he says. "When you stumble on a beautiful theory you have the same emotional reaction that you feel in front of a piece of art."

It's not that I don't know what he is talking about; I don't know why it matters. I doubt my sense of beauty is a reliable guide to uncovering fundamental laws of nature, laws that dictate the behavior of entities that I have no direct sensory awareness of, never had, and never will have. For it to be hardwired in my brain, it ought to have been beneficial during natural selection. But what evolutionary advantage has there ever been to understanding quantum gravity?

And while creating works of beauty is a venerable craft, science isn't art. We don't seek theories to evoke emotional reactions; we seek explanations for what we observe. Science is an organized enterprise to overcome the shortcomings of human cognition and to avoid the fallacies of intuition. Science isn't about emotion—it's about numbers and equations, data and graphs, facts and logic.

I think I wanted him to prove me wrong.

When I ask Gian what he makes of the recent LHC data, he says: "We are so confused." Finally, something I understand.

Failure

In the first years of its operation, the LHC dutifully delivered a particle called the Higgs boson, the existence of which had been predicted back in the 1960s. My colleagues and I had high hopes that this billion-dollar project would do more than just confirm what nobody doubted. We had found some promising cracks in the foundations that convinced us the LHC would also create other, so far undiscovered particles. We were wrong. The LHC hasn't seen anything that would support our newly invented laws of nature.

Our friends in astrophysics haven't fared much better. In the 1930s they had discovered that galaxy clusters contain a lot more mass than all visible matter combined can possibly account for. Even allowing for large uncertainty in the data, a new type of "dark matter" is needed to explain the observations. Evidence for the gravitational pull of dark matter has piled up, so we are sure it is there. What it is made of, however, has remained a mystery. Astrophysicists believe it is some type of particle not present here on Earth, one that neither absorbs nor emits light. They thought up new laws of nature, unconfirmed theories, to guide the construction of detectors meant to test their ideas. Starting in the 1980s, dozens of experimental crews have been hunting for these hypothetical dark matter particles. They haven't found them. The new theories have remained unconfirmed.

It looks similarly bleak in cosmology, where physicists try in vain to understand what makes the universe expand faster and faster, an observation attributed to "dark energy." They can mathematically show that this strange substrate is nothing but the energy carried by empty space, and yet they cannot calculate the amount of energy. It's one of the cracks in the foundations that physicists attempt to peer through, but so far they have failed to see anything in support of the new theories they designed to explain dark energy.

Meanwhile, in the field of quantum foundations, our colleagues want to improve a theory that has no shortcomings whatsoever. They act based on the conviction that something is wrong with mathematical structures that don't correspond to measurable entities. It irks them that, as Richard Feynman, Niels Bohr, and other heroes of last century's physics complained, "nobody understands quantum mechanics." Researchers in quantum foundations want to invent better theories, believing, as everyone else does, they are on the right crack. Alas, all experiments have upheld the predictions of the not-understandable theory from the last century. And the new theories? They are still untested speculations.

An enormous amount of effort went into these failed attempts to find new laws of nature. But for more than thirty years now we have not been able to improve the foundations of physics.

ooOo

SO YOU want to know what holds the world together, how the universe was made, and what rules our existence goes by? The closest you will get to an answer is following the trail of facts down into the basement of science. Follow it until facts get sparse and your onward journey is blocked by theoreticians arguing whose theory is prettier. That's when you know you've reached the foundations.

The foundations of physics are those ingredients of our theories that cannot, for all we presently know, be derived from anything simpler. At this bottommost level we presently have space, time, and twenty-five particles, together with the equations that encode their behavior. The subjects of my research area, then, are particles that move through space and time, occasionally hitting each other or forming composites. Don't think of them as little balls; they are not, because of quantum mechanics (more about that later). Better think of them as clouds that can take on any shape.

In the foundations of physics we deal only with particles that cannot be further decomposed; we call them "elementary particles." For all we presently know, they have no substructure. But the elementary particles can combine to make up atoms, molecules, proteins—and thereby create the enormous variety of structure we see around us.

It's these twenty-five particles that you, I, and everything else in the universe are made of.

But the particles themselves aren't all that interesting. What is interesting are the relations between them, the principles that determine their interaction, the structure of the laws that gave birth to the universe and enabled our existence. In our game, it's the rules we care about, not the pieces. And the most important lesson we have learned is that nature plays by the rules of mathematics.

Made of Math

In physics, theories are made of math. We don't use math because we want to scare away those not familiar with differential geometry and graded Lie algebras; we use it because we are fools. Math keeps us honest—it prevents us from lying to ourselves and to each other. You can be wrong with math, but you can't lie.

Our task as theoretical physicists is to develop the mathematics to either describe existing observations, or to make predictions that guide experimental strategies. Using mathematics in theory development enforces logical rigor and internal consistency; it ensures that theories are unambiguous and conclusions are reproducible.

The success of math in physics has been tremendous, and consequently this quality standard is now rigorously enforced. The theories we build today are sets of assumptions—mathematical relations or definitions—together with interpretations that connect the math with real-world observables.

But we don't develop theories by writing down assumptions and then derive observable consequences in a sequence of theorems and proofs. In physics, theories almost always start out as loose patchworks of ideas. Cleaning up the mess that physicists generate in theory development, and finding a neat set of assumptions from which the whole theory can be derived, is often left to our colleagues in mathematical physics—a branch of mathematics, not of physics.

For the most part, physicists and mathematicians have settled on a fine division of labor in which the former complain about the finickiness of the latter, and the latter complain about the sloppiness of the former. On both sides, though, we are crucially aware that progress in one field drives progress in the other. From probability theory to chaos theory to the quantum field theories at the base of modern particle physics, math and physics have always proceeded hand in hand.

But physics isn't math. Besides being internally consistent (not giving rise to conclusions that contradict each other) a successful theory must also be consistent with observation (not be in contradiction with the data). In my area of physics, where we deal with the most fundamental questions, this is a stringent demand. There is so much existing data that doing all the necessary calculations for newly proposed theories simply isn't feasible. It is also unnecessary because there is a shortcut: We first demonstrate that a new theory agrees with the well-confirmed old theories to within measurement precision, thus reproducing the old theory's achievements. Then we only have to add calculations for what more the new theory can explain.

Demonstrating that a new theory reproduces all the achievements of successful old theories can be extremely difficult. This is because a new theory might use an entirely different mathematical framework that looks nothing like that of the old theory. Finding a way to show that both nevertheless arrive at the same predictions for already-made observations often requires finding a suitable way to reformulate the new theory. This is straightforward in cases where the new theory directly employs the math of the old one, but it can be a big hurdle with entirely new frameworks.

Einstein, for example, struggled for years to prove that general relativity, his new theory of gravity, would reproduce the successes of the predecessor, Newtonian gravity. The problem wasn't that he had the wrong theory; the problem was that he didn't know how to find Newton's gravitational potential in his own theory. Einstein had all the math right, but the identification with the real world was missing. Only after several wrong attempts did he hit on the right way to do it. Having the right math is only part of having the right theory.

There are other reasons we use math in physics. Besides keeping us honest, math is also the most economical and unambiguous terminology that we know of. Language is malleable; it depends on context and interpretation. But math doesn't care about culture or history. If a thousand people read a book, they read a thousand different books. But if a thousand people read an equation, they read the same equation.

The main reason we use math in physics, however, is because we can.

Physics Envy

While logical consistency is always a requirement for a scientific theory, not all disciplines lend themselves to mathematical modeling—using a language so rigorous doesn't make sense if the data don't match the rigor. And of all the scientific disciplines, physics deals with the simplest of systems, making it ideally suited for mathematical modeling.

In physics, the subjects of study are highly reproducible. We understand well how to control experimental environments and which effects can be neglected without sacrificing accuracy. Results in psychology, for example, are hard to reproduce because no two people are alike, and it is rarely known exactly which human quirks might play a role. But that's a problem we don't have in physics. Helium atoms don't get hungry and are just as well-tempered on Monday as on Friday.

This precision is what makes physics so successful, but also what makes it so difficult. To the uninitiated, the many equations might appear inaccessible, but handling them is a matter of education and habituation. Understanding the math is not what makes physics difficult. The real difficulty is finding the right math. You can't just take anything that looks like math and call it a theory. It's the requirement that a new theory needs to be consistent, both internally consistent and consistent with experiment—with each and every experiment—that makes it so difficult.

Theoretical physics is a highly developed discipline. The theories that we work with today have stood up to a great many experimental tests. And every time the theories passed another test, it has become a little more difficult to improve anything about them. A new theory needs to accommodate all of the present theories' success and still be a little better.

As long as physicists developed theories to explain existing or upcoming experiments, success meant getting the right numbers with the least amount of effort. But the more observations our theories could describe, the more difficult it became to test a proposed improvement. It took twenty-five years from the prediction of the neutrino to its detection, almost fifty years to confirm the Higgs boson, a hundred years to directly detect gravitational waves. Now the time it takes to test a new fundamental law of nature can be longer than a scientist's full career. This forces theorists to draw upon criteria other than empirical adequacy to decide which research avenues to pursue. Aesthetic appeal is one of them.

In our search for new ideas, beauty plays many roles. It's a guide, a reward, a motivation. It is also a systematic bias.

Invisible Friends

The movers have picked up my boxes, most of which I never bothered to unpack, knowing I wouldn't stay here. Echoes of past moves return from empty cabinets. I call my friend and colleague Michael Krämer, professor of physics in Aachen, Germany.

Michael works on supersymmetry, "susy" for short. Susy predicts a large number of still undiscovered elementary particles, a partner for each of the already known particles and a few more. Among the proposed new laws of nature, susy is presently the most popular one. Thousands of my colleagues bet their careers on it. But so far, none of those extra particles have been seen.

"I think I started working on susy because that's what people worked on when I was a student, in the mid- to late nineties," says Michael.

The mathematics of susy is very similar to that of already established theories, and the standard physics curriculum is good preparation for students to work on susy. "It's a well-defined framework; it was easy," says Michael. It was a good choice. Michael received tenure in 2004 and now heads the research group New Physics at the Large Hadron Collider, funded by the German Research Foundation.

"I also like symmetries. That made it attractive for me."

ooo

As I've noted, on our quest to understand what the world is made of, we have found twenty-five different elementary particles. Supersymmetry completes this collection with a set of still undiscovered partner particles, one for each of the known particles, and some additional ones. This supersymmetric completion is appealing because the known particles are of two different types, fermions and bosons (named after Enrico Fermi and Satyendra Bose, respectively), and supersymmetry explains how these two types belong together.

Fermions are extreme individuals. No matter how hard you try, you will not get two of them to do the same thing in the same place—there must always be a difference between them. Bosons, on the other hand, have no such constraint and are happy to join each other in a common dance. This is why electrons, which are fermions, sit on separate shells around atomic nuclei. If they were bosons, they would instead sit together on the same shell, leaving the universe without chemistry—and without chemists, as our own existence rests on the little fermions' refusal to share space.

Supersymmetry postulates that the laws of nature remain the same when bosons are exchanged with fermions. This means that every known boson must have a fermionic partner, and every known fermion must have a bosonic partner. But besides differing in their fermionic or bosonic affiliation, partner particles must be identical.

Since none of the already known particles match as required, we have concluded there are no supersymmetric pairs among them. Instead, new particles must be waiting to be discovered. It's like we have a collection of odd pots and lids and are convinced that certainly the matching pieces must be around somewhere.

Unfortunately, the equations of supersymmetry do not tell us what the masses of the susy partners are. Since it takes more energy to produce heavier particles, a particle is more difficult to find when its mass is larger. All we have learned so far is that the superpartners, if they exist, are so heavy that the energy of our experiments isn't yet large enough to create them.

Supersymmetry has much going for it. Besides revealing that bosons and fermions are two sides of the same coin, susy also aids in the unification of fundamental forces and has the potential to explain several numerical coincidences. Moreover, some of the supersymmetric particles have just the right properties to make up dark matter. I'll tell you more about that in the later chapters.

ꝏꝏ

SUPERSYMMETRY FITS so snugly with the existing theories that many physicists are convinced it must be right. "Despite the efforts of many hundreds of physicists conducting experiments in search of these particles, no superpartners have ever been observed or detected," writes Fermilab physicist Dan Hooper. Yet "this has had little effect in deterring the theoretical physicists who passionately expect nature to be formulated this way—to be supersymmetric. To many of these scientists, the ideas behind supersymmetry are simply too beautiful and too elegant not to be part of our universe. They solve too many problems and fit into our world too naturally. To these true believers, the superpartner particles simply must exist."[2]

Hooper isn't the only one to emphasize the strength of this conviction. "For many theoretical physicists, it is hard to believe that supersymmetry does not play a role somewhere in nature," notes physicist Jeff Forshaw.[3] And in a 2014 *Scientific American* article titled "Supersymmetry and the Crisis in Physics," particle physicists Maria Spiropulu and Joseph Lykken support their hope that evidence will come in, eventually, with the assertion that "it is not an exaggeration to say that most of the world's particle physicists believe that supersymmetry *must* be true" (their emphasis).[4]

It adds to susy's appeal that a symmetry relating bosons and fermions was long thought impossible because a mathematical proof seemed to forbid it.[5] But no proof is better than its assumptions. It turned out that if the proof's assumptions are relaxed, supersymmetry instead is the largest possible symmetry that can be accommodated in the existing theories.[6] And how could nature not make use of such a beautiful idea?

ꝏꝏ

"FOR ME the most beautiful aspect of susy was always that it was the biggest kind of symmetry," recalls Michael. "I found this appealing. When I learned about this exception I thought, 'Oh, this is interesting,' because to me it seemed that this idea—you impose symmetries and you find the right laws of nature, even if you don't understand exactly why it works—seems like a powerful principle. So it seemed to me worthwhile to pursue this."

When I was a student, in the late 1990s, the simplest susy models had already run into conflict with data and the process of designing more complicated but still viable models had begun.[7] To me it looked like a field where nothing new could be said without first detecting the predicted particles. I decided to stay away from susy until that happened.

It hasn't happened. No evidence for susy was found at the Large Electron Positron (LEP) collider, which ran until 2000. Neither was anything found at the Tevatron, a collider that reached higher energies than LEP and that ran until 2011. The even more powerful LHC, which reused LEP's tunnel, has been running since 2008, but susy hasn't shown up.

Still, I worry that I made a big mistake not going into the field that so many of my colleagues regarded, and continue to regard, as so promising.

For many years, the lore was that something new has to appear at the LHC because otherwise the best existing description of particle physics—the standard model—would not be natural according to the measures introduced, among others, by Gian Francesco Giudice. These mathematical formulae to measure naturalness rest on the belief that a theory with very large or very small numbers isn't pretty.

We will explore throughout the rest of this book whether this belief is justified. For now it suffices to say it's widespread. In a 2008 paper, Giudice explained: "The concept of naturalness... developed through a 'collective motion' of the community which increasingly emphasized their relevance to the existence of physics beyond the Standard Model."[8] And the more they studied naturalness, the more they became convinced that to avoid ugly numerical coincidences new discoveries had to be made soon.

"In hindsight, it is surprising how much emphasis was put on this naturalness argument," says Michael. "If I look back, people repeated the same argument, again and again, not really reflecting on it. They were saying the same thing, saying the same thing, for ten years. It is really surprising that this was the main driver for so much of model building. Looking back, I find this strange. I still think naturalness is appealing, but I'm not convinced anymore that this points to new physics at the LHC."

The LHC finished its first run in February 2013, then shut down for an upgrade. The second run at higher energies started in April 2015. Now it is October 2015, and in the coming months we expect to see preliminary results of run two.

"You should talk to Arkani-Hamed," Michael says. "He is a naturalness supporter—a very interesting guy. He is really influential, especially in the US—it's amazing. He works on something for a while and gathers followers, and then he moves to something else the next year. Ten years ago, he worked on this model with natural susy and he talked about it so convincingly that everybody started looking into this. And then two years later he writes this paper on unnatural susy!"

Nima Arkani-Hamed made his name in the late 1990s for proposing, together with Savas Dimopoulos and Gia Dvali, that our universe might have additional dimensions, rolled up to small radii but still large enough to be testable with particle accelerators.[9] The idea that additional dimensions exist is not new—it dates back to the 1920s.[10] The genius of Arkani-Hamed and collaborators was to propose that these dimensions are so large they might become testable soon, a suggestion that inspired thousands of physicists to calculate and publish further details. The argument for why the LHC should reveal the extra dimensions was naturalness. "Naturalness requires that the migration into the extra dimensions cannot be postponed much beyond the TeV scale," the authors argued in their first work on what is now known, after their initials, as the ADD model.* To date, the paper has been cited more than five thousand times. That makes it one of the most-cited papers in physics, ever.

In 2002, after I got stuck with a self-chosen PhD topic about a variant of the 1920s version of extra dimensions, my supervisor convinced me that I had better switch to its modern incarnation. And so I too wrote some papers on testing extra dimensions at the LHC. But the LHC hasn't seen any evidence for extra dimensions. I began questioning arguments from naturalness. Nima Arkani-Hamed moved on from large extra-dimensions to susy and is now professor of physics at the Institute for Advanced Studies in Princeton.

I make a mental note to talk to Nima.

"He's much harder to get than I am, of course. I don't think he replies to emails that easily," Michael tells me. "He is driving the whole US particle physics landscape. And he has this argument that we need a 100 TeV collider to test naturalness. And now maybe the Chinese will build his collider—who knows!"

As it becomes increasingly clear that the LHC will not deliver the awaited evidence for prettier laws of nature, particle physicists once again shift hopes to the next bigger collider. Nima is one of the main advocates for building a new circular particle accelerator in China.

---

* The abbreviation *eV* stands for "electron volt" and is a measure of energy. A TeV is 10^{12}, or a trillion, eV. The LHC can maximally deliver about 14 TeV. Hence the LHC is said to be "testing the TeV scale."

But regardless of what else might be discovered at higher energies, that the LHC so far hasn't found any new elementary particles means that the correct theory is, by physicists' standards, unnatural. We have indeed maneuvered ourselves into an oxymoronic situation in which, according to our own beauty requirements, nature itself is unnatural.

"Am I worried? I don't know. I'm confused," says Michael, "I'm honestly confused. Before the LHC, I thought something must happen. But now? I'm confused." It sounds familiar.

IN BRIEF

- Physicists use a lot of math and are really proud that it works so well.
- But physics isn't math, and theory development needs data for guidance.
- In some areas of physics there hasn't been new data for decades.
- In the absence of guidance from experiments, theorists use aesthetic criteria.
- They get confused if that doesn't work.

2
What a Wonderful World

In which I read a lot of books about dead people and find that everyone likes pretty ideas but that pretty ideas sometimes work badly. At a conference I begin to worry that physicists are about to discard the scientific method.

Where We Come From

While I was in school I hated history, but since then I have come to recognize the usefulness of quoting dead people to support my convictions. I won't even pretend to give you a historical account of the role of beauty in science, because really I am more interested in the future than in the past, and also because others have been to the scene before.[1] But if we are to see how physics has changed, I have to tell you how it used to be.

Until the end of the nineteenth century it was quite common for scientists to view nature's beauty as a sign of divinity. As they sought—and found—explanations that formerly had been territory of the church, the inexplicable harmony revealed by the laws of nature reassured the religious that science posed no risk for the supernatural.

Around the turn of the century, when science separated from religion and became more professionalized, its practitioners ceased to assign the beauty of natural law to godly influence. They marveled over harmony in the rules that govern the universe but left the interpretation of that harmony open, or at least marked their belief as personal opinion.

In the twentieth century, aesthetic appeal morphed from a bonus of scientific theories to a guide in their construction until, finally, aesthetic principles turned into mathematical requirements. Today we don't reflect on arguments from beauty anymore—their nonscientific origins have gotten "lost in math."

∞∞

AMONG THE first to formulate quantitative laws of nature was the German mathematician and astronomer Johannes Kepler (1571–1630), whose work was strongly influenced by his religious belief. Kepler had a model for the solar system in which the then-known planets—Mercury, Venus, Earth, Mars, Jupiter, and Saturn—moved in circular orbits around the Sun. The radii of their orbits were determined by regular polyhedra—the Platonic solids—stacked inside each other, and the distances between the planets thus obtained fit well with observations. It was a pretty idea: "It is absolutely necessary that the work of such a perfect creator should be of the greatest beauty," Kepler opined.

With the help of tables that detailed the planets' exact positions, Kepler later convinced himself that his model was wrong, and concluded that the planets move in ellipses, not circles, around the Sun. His new idea was promptly met with disapproval; he had failed to meet the aesthetic standard of the time.

He received criticism in particular from Galileo Galilei (1564–1641), who believed that "only circular motion can naturally suit bodies which are integral parts of the universe as constituted in the best arrangement."[2] Another astronomer, David Fabricius (1564–1617), complained that "with your ellipse you abolish the circularity and uniformity of the motions, which appears to me the more absurd the more profoundly I think about it." Fabricius, as many at the time, preferred to amend the planetary orbits by adding "epicycles," which were smaller circular motions around the already circular orbits. "If you could only preserve the perfect circular orbit and justify your elliptic orbit by another little epicycle," Fabricius wrote to Kepler, "it would be much better."[3]

But Kepler was right. The planets do move in ellipses around the Sun.

After evidence forced him to give up the beautiful polyhedra, Kepler, in later life, became convinced that the planets play music along their paths. In his 1619 book *Harmony of the World* he derived the planet's tunes and concluded that "the Earth sings Mi-Fa-Mi." It wasn't his best work. But Kepler's analysis of the planetary orbits laid a basis for the later studies of Isaac Newton (1643–1727), the first scientist to rigorously use mathematics.

Newton believed in the existence of a god whose influence he saw in the rules that nature obeyed: "This most beautiful system of the Sun, planets, and comets, could only proceed from the counsel and dominion of an intelligent Being," he wrote in 1726.[4] Since their inception, Newton's laws of motion and gravitation have been radically overhauled, but they remain valid today as approximations.

Newton and his contemporaries had no qualms about combining religion with science—back then this was generally accepted procedure. The most inclusive of them all might have been Gottfried Wilhelm Leibniz (1646–1716), who developed calculus around the same time, but independently of Newton. Leibniz believed the world we inhabit is "the best of all possible worlds" and all existing evil is necessary. He argued that the shortcomings of the world are "based upon the too slight acquaintance which we have with the general harmony of the universe and with the hidden reasons for God's conduct."[5] In other words, according to Leibniz, the ugly is ugly because we don't understand what beauty is.

Leibniz's argument, as much as philosophers and theologians like to argue about it, is useless without defining what "best" even means. But the underlying idea that our universe is optimal in some sense gained a foothold in science and stomped through the centuries. Once expressed mathematically, it grew into a giant on whose shoulders all modern physical theories stand.[6] Contemporary theories merely differ in the way in which they require a system to behave in the "best" way. Einstein's theory of general relativity, for example, can be derived by requiring the curvature of space-time to be as small as possible; similar methods exist for the other interactions. Still, physicists today struggle to find an overarching principle according to which our universe is the "best"—a problem we will come back to later.

How We Got Here

As the centuries passed and mathematics became more powerful, references to God in physics slowly faded away, or combined with the laws of nature themselves. At the end of the nineteenth century, Max Planck (1858–1947) believed that "the holiness of the unintelligible Godhead is conveyed by the holiness of symbols." Then, as the nineteenth century turned into the twentieth, beauty gradually morphed into a guiding principle for theoretical physicists, a transition that was solidified with the development of the standard model.

Hermann Weyl (1885–1955), a mathematician who made important contributions to physics, was rather unapologetic about his not-so scientific methods: "My work always tries to unite the true with the beautiful; but when I had to choose one or the other, I usually chose the beautiful."[7] The astrophysicist Edward Arthur Milne (1896–1950), influential during the development of general relativity, regarded "beauty as a road to knowledge, or rather as being the only knowledge worth having." In a 1922 talk at the Cambridge University Natural Science Club, he complained about the profusion of ugly research:

> *One only has to look through the back files of scientific periodicals, say of 50 years back, to come across dozens of papers which have served no purpose in the extension of scientific knowledge and which never could have done, mere fungi on the trunk of the scientific tree, and like fungi,*

> *constantly reappearing if swept away.... [But if a paper] evokes in us those emotions which we associate with beauty no further justification is needed; it is not a fungus but a blossom; it is a terminus of science, the end of a line of inquiry in which science* has *reached its ultimate goal. It is the ugly papers which require justification.*[8]

Paul Dirac (1902–1984), a Nobel laureate who has an equation named after him, went a step further and spelled out instructions: "The research worker, in his efforts to express the fundamental laws of Nature in mathematical form, should strive mainly for mathematical beauty."[9] On another occasion, when asked to summarize his philosophy of physics, Dirac took to the blackboard and wrote "PHYSICAL LAWS SHOULD HAVE MATHEMATICAL BEAUTY."[10] The historian Helge Kragh concluded his biography of Dirac with the observation that "after 1935 [Dirac] largely failed to produce physics of lasting value. It is not irrelevant to point out that the principle of mathematical beauty governed his thinking only during the later period."[11]

Albert Einstein, who really needs no introduction, worked himself into a state in which he believed that thought alone can reveal the laws of nature: "I am convinced that we can discover by means of purely mathematical constructions the concepts and the laws connecting them with each other, which furnish the key to the understanding of natural phenomena.... In a certain sense, therefore, I hold it true that pure thought can grasp reality, as the ancients dreamed."[12] To be fair to the man, he did in other instances emphasize the need for observation.

Jules Henri Poincaré, who made many contributions to both math and physics but is perhaps best-known for his discovery of deterministic chaos, praised the practical use of beauty: "Thus we see that care for the beautiful leads us to the same selection as care for the useful."[13] Poincaré considered "economy of thought" (*Denkökonomie*—a term coined by Ernst Mach) to be "a source of beauty as well as a practical advantage." The human aesthetic sense, he argued, "plays the part of the delicate sieve" that helps the researcher to develop a good theory, and "this harmony is at once a satisfaction of our aesthetic requirements and an assistance to the mind which it supports and guides."[14]

And Werner Heisenberg, one of the founders of quantum mechanics, boldly believed that beauty has a grasp on truth: "If nature leads us to mathematical forms of great simplicity and beauty we cannot help thinking that they are 'true,' that they reveal a genuine feature of nature."[15] As his wife recalls:

> *One moonlit night, we walked over the Hainberg Mountain, and he was completely enthralled by the visions he had, trying to explain his newest discovery to me. He talked about the miracle of symmetry as the original archetype of creation, about harmony, about the beauty of simplicity, and its inner truth.*[16]

Beware the moonlight walks with theoretical physicists—sometimes enthusiasm gets the better of us.

What We Are Made Of

When I was a teenager, in the 1980s, there weren't many popular science books about contemporary theoretical physics or, heaven forbid, mathematics. Dead people's biographies were the place to look. Browsing through books in the library, I pictured myself a theoretical physicist, puffing on a pipe while sitting in a leather armchair, thinking big thoughts while absentmindedly stroking my beard. Something seemed wrong with that picture. But the message that math plus thought can decode nature made a deep impression on me. If this was a skill that could be learned, it was what I wanted to learn.

One of the few popular science books that covered modern physics in the 1980s was Anthony Zee's *Fearful Symmetry*.[17] Zee, who was then and still is a professor of physics at the University of California, Santa Barbara, wrote: "My colleagues and I, we are the intellectual descendants of Albert Einstein; we like to think that we too search for beauty." And he laid out the program: "In this century physicists have become increasingly ambitious. . . . No longer content to explain this phenomenon or that, they have become imbued with the faith that Nature has an underlying design of beautiful simplicity."

Not only have they become "imbued with faith" in beauty, but they have found means to express their faith in mathematical form: "Physicists developed the notion of symmetry as an objective criterion in judging Nature's design," Zee wrote. "Given two theories, physicists feel that the more symmetrical one, generally, is the more beautiful. When the beholder is a physicist, beauty *means* symmetry."

ꝏꝏ

FOR THE physicist, a symmetry is an organizing principle that avoids unnecessary repetition. Any type of pattern, likeness, or order can be mathematically captured as an expression of symmetry. The presence of a symmetry always reveals a redundancy and allows simplification. Hence, symmetries explain more with less.

For example, rather than telling you today's sky looks blue in the west and the east and the north and the south and the southwest, and so on, I can just say it looks blue in every direction. This independence on the direction is a rotational symmetry, and it makes it sufficient to spell out how a system looks in one direction, followed by saying it's the same in all other directions. The benefit is fewer words or, in our theories, fewer equations.

The symmetries that physicists deal with are more abstract versions of this simple example, like rotations among multiple axes in internal mathematical spaces. But it always works the same way: find a transformation under which the laws of nature remain invariant and you've found a symmetry. Such a symmetry transformation may be anything for which you can write down an unambiguous procedure—a shift, a rotation, a flip, or really any other operation that you can think of. If this operation does not make a difference to the laws of nature, you have found a symmetry. With that, you save the effort of having to explain the changes the operation leads to; instead, you can just state there are no changes. It's Mach's "economy of thought."

In physics we use many different types of symmetries, but they have one thing in common: they are potent unifying principles because they explain how things that once appeared very different actually belong together, connected by a symmetry transformation. Often, however, it isn't easy to find the correct symmetry to simplify large stacks of data.

The most stunning success of symmetry principles might have been the development of the quark model. Since the advent of particle colliders in the 1930s, physicists had been slamming particles together at ever higher energies. By the mid-1940s, they were reaching energies that could probe the structure of the atomic core, and the number of particles began blowing up. First there were the charged pions and kaons. Then came the neutral pion and the neutral kaon, the first delta resonances, a particle dubbed lambda, the charged sigmas, the rhos, an omega, the eta, the K-star, the phi meson—and that was only the beginning. Enrico Fermi, when asked by Leon Lederman what he thought about the recent discovery of a particle named the K-zero-two, said: "Young man, if I could remember the names of these particles I would have been a botanist."[18]

Altogether, physicists detected hundreds of particles, all of which were unstable and decayed quickly. These particles seemed to have no apparent relation to each other, quite contrary to the physicists' hope that the laws of nature would get simpler for the more fundamental constituents of matter. By the 1960s, accommodating this "particle zoo" in a comprehensive theory had become a major research agenda.

One of the most popular approaches at the time was to just forgo the desire for explanation, and to collect the particles' properties in a big table—the scattering matrix or S-matrix—which was the very opposite of beauty and economy. Then came Murray Gell-Mann. He identified the correct properties of the particles—called hypercharge and isospin—and it turned out all the particles could be classified by symmetric patterns known as multiplets.

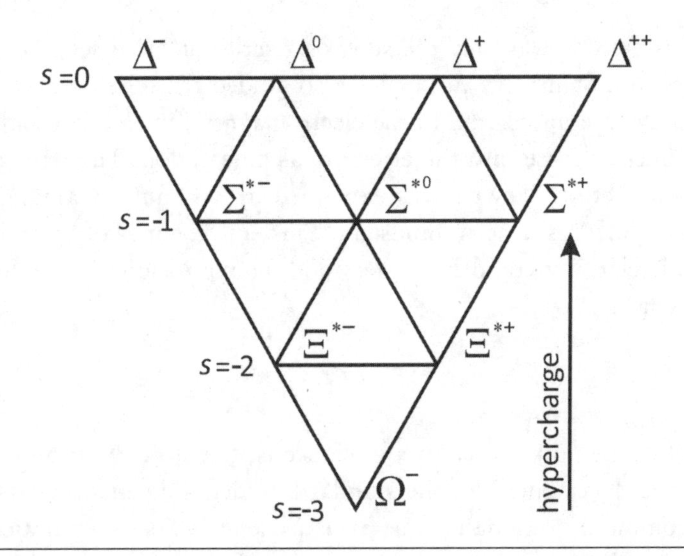

FIGURE 1. The baryon decouplet is an example of the use of symmetries in theoretical physics. Gell-Mann used its incompleteness to predict the omega minus (Ω^-), the particle at the bottom tip.

Later it became clear that the multiplets' regularities meant the observed particles were composed of smaller entities that—for reasons not well understood at the time—had never been detected in isolation. Gell-Mann called the smaller constituents "quarks."[19] The light composites, called mesons, are made of two quarks, and the heavier composites, called baryons, are made of three quarks. (All mesons are unstable. The baryons include neutrons and protons, which build atomic nuclei.)

The symmetry in the resulting patterns, once revealed, is apparent to the eye (see Figure 1). Remarkably, when Gell-Mann proposed this idea, several of the multiplets were still incomplete. The symmetry requirement therefore prompted him to predict the particles necessary to fill in the pattern, in particular a baryon known as omega minus. It was later found with the properties that Gell-Mann had calculated, and he was awarded the Nobel Prize in 1969. Beauty had won over the ugly, postmodern S-matrix approach.

This episode was only the start of a series of successes scored by symmetries. Symmetry principles also guided the development and, eventually, the unification of the electromagnetic interaction with the weak nuclear force into the electroweak interaction. The strong nuclear interaction likewise was explained by a symmetry among elementary particles. And in hindsight, Einstein's theories of special and general relativity could be understood as expressions of symmetry requirements.

∞∞

THE MODERN faith in beauty's guidance is, therefore, built on its use in the development of the standard model and general relativity; it is commonly rationalized as an experience value: they noticed it works, and it seems only prudent to continue using it. Gell-Mann himself relates that "in fundamental physics a beautiful or elegant theory is more likely to be right than a theory that is inelegant."[20] Lederman, the young man who asked Fermi about the K-zero-two, went on to win a Nobel Prize as well, and he too became a beauty convert: "We believe that nature is best described in equations that are as simple, beautiful, compact and universal as possible."[21]

Steven Weinberg, who was awarded a Nobel Prize for unifying the electromagnetic and weak interaction, likes to make an analogy with horse breeding: "[The horse breeder] looks at a horse and says 'That's a beautiful horse.' While he or she may be expressing a purely aesthetic emotion, I think there's more to it than that. The horse breeder has seen lots of horses, and from experience with horses knows that that's the kind of horse that wins races."[22]

But just as experience with horses doesn't help when building a race car, experience with last century's theories might not be of much help conceiving better ones. And without justification from experience beauty remains as subjective as ever. This apparent clash with the scientific method is acknowledged by today's physicists, but using aesthetic criteria has nevertheless become widely accepted practice.

And the more removed from experimental test a subject area is, the more relevant the aesthetic appeal of its theories.

In foundational physics, which is as far from experimental test as science can be while still being science, the influence of beauty judgments is particularly pronounced. Few of my colleagues even bother denying that they pay more attention to theories they consider pretty. Their archetypal caution against subjective assessments is inevitably followed by a "but" and a referral to common practice.

Frank Wilczek, for example, who, together with David Gross and Hugh David Politzer was awarded the 2004 Nobel Prize for their work on the strong nuclear force, writes in his book *A Beautiful Question* that "our fundamental sense of beauty is not in any very direct way adapted to Nature's fundamental workings." But, "having tasted beauty at the heart of the world, we hunger for more. In this quest, I think, there is no more promising guide than beauty itself."[23]

Gerard 't Hooft, who first formulated a mathematical criterion for naturalness that now drives much of the research in theoretical particle physics (and who has also won a Nobel Prize), warns: "Beauty is a dangerous concept, because it can always mislead people. If you have a theory that is more beautiful than you expected at first, that's an indication that you might be correct, that you might be right. But it's not a guarantee at all. In your eyes a theory might be beautiful, but it might just be wrong. There's nothing you can do about it." But: "Certainly when we read about new theories and we see how beautiful and simple they are, then they have a big advantage. We believe such theories have much more chance to be successful."[24]

In his best-selling book *The Elegant Universe*, string theorist Brian Greene (who has not won a Nobel Prize) assures the reader: "Aesthetic judgments do not arbitrate scientific discourse." Then he continues, "But it is certainly the case that some decisions made by theoretical physicists are founded upon an aesthetic sense—a sense of which theories have an elegance and beauty of structure on par with the world we experience.... So far, this approach has provided a powerful and insightful guide."[25]

Abstract mathematics is hard to convey, and this human quest for beauty could be dismissed as a marketing aid for popular science books. But the popular books do more than just make a difficult subject approachable—they reveal how theoretical physicists think and work.

Where Beauty Lies

Last century's triumphs are still fresh in the minds of researchers now nearing retirement, and their emphasis on beauty has greatly influenced the following generation—my generation, the unsuccessful generation. We work with the now formalized aesthetic ideals of the past: symmetry, unification, and naturalness.

It seems only reasonable to draw upon past experience and try what worked before. Indeed, we'd be stupid not to take advice from those who came before us. But we'd also be stupid to get stuck on it. And I'm wary, getting warier with each null result. Beauty is a treacherous guide, and it has led physicists astray many times before.

ooo

> *That these interrelations display, in all their mathematical abstraction, an incredible degree of simplicity, is a gift we can only accept humbly. Not even Plato could have believed them to be so beautiful. For these interrelationships cannot be invented; they have been there since the creation of the world.*

So wrote Heisenberg in a 1958 letter to his sister Edith.[26] The beautiful interrelationships that Heisenberg refers to here, however, are not those of his theory of quantum mechanics. No, in this period of his life he attempted, and failed, to develop a unified theory, now little more than a side note in the history books.

And when we look at Heisenberg's ideas that were successful, we find that his scientific works didn't exactly register as marvels of beauty. His contemporary Erwin Schrödinger commented: "I knew of [Heisenberg's] theory, of course, but I felt discouraged, not to say repelled, by the methods of transcendental algebra, which appeared difficult to me, and by the lack of visualizability."[27]

Not that Heisenberg was any nicer about Schrödinger's ideas. In a letter to Wolfgang Pauli he wrote: "The more I think about the physical portion of the Schrödinger theory, the more repulsive I find it. What Schrödinger writes about visualizability of his theory... I think it's crap."[28] In the end, both Heisenberg's and Schrödinger's approaches turned out to be part of the same theory.

The advent of quantum mechanics wasn't the only beauty fail in physics. The Platonic solids that Kepler used to calculate planetary orbits, which we heard about earlier, may be the best-known example for the conflict between aesthetic ideals and facts. A more recent case, dating back to first half of the twentieth century, is the steady state model for the universe.

In 1927, Georges Lemaître found a solution to the equations of general relativity which lead him to propose that a matter-filled universe like ours expands. He concluded that the universe must have had a beginning, the "big bang." Einstein, when first confronted with this solution, informed Lemaître that he found the idea "abominable."[29] He had instead introduced an additional term in his equations—the cosmological constant—to force the universe into a static configuration.

In 1930, however, Arthur Eddington, who had been instrumental in organizing the first experimental test of general relativity, showed that Einstein's solution with the cosmological constant is unstable: even the smallest shift in matter distribution would make it collapse or expand. This instability, together with observations by Edwin Hubble that supported Lemaître's idea, led Einstein in 1931 to also adopt the expanding universe.

Still, for many decades after this, cosmology remained starved of data and offered a playground for philosophical and aesthetic debate. Arthur Eddington in particular held on to Einstein's static universe because he believed the cosmological constant represented a new force of nature. He dismissed Lemaître's idea on the ground that "the notion of a beginning of the world is repugnant to me."

In his late years, Eddington developed a "fundamental theory" that was supposed to join the static cosmology with quantum theory. In this attempt he drifted off into his own cosmos: "In science we sometimes have convictions as to the right solution of a problem which we cherish but cannot justify; we are influenced by some innate sense of the fitness of things." Because of the increasing tension with data, Eddington's fundamental theory was not further pursued after his death in 1944.

The idea of an unchanging universe remained popular, however. To make it compatible with the observed expansion, Hermann Bondi, Thomas Gold, and Fred Hoyle proposed in 1948 that matter was continuously produced between the galaxies. In this case we would be living in an eternally expanding universe, but one without beginning and without end.

Fred Hoyle's motivations in particular were based on aesthetic grounds. He made fun of Lemaître by calling him "the big bang man" and admitted, "I have an aesthetic bias against the big bang."[30] In 1992, when the American George Smoot announced the measurement of temperature fluctuations in the cosmic background radiation that spoke against the steady state idea, Hoyle (who died in 2001) refused to accept it. He revised his model to a "quasi-steady state cosmology" to accommodate the data. His explanation for the success of Lemaitre's idea was that "the reason why scientists like the 'big bang' is because they are overshadowed by the Book of Genesis."[31]

Aesthetic ideals also gave rise to what may be the strangest episode in the history of physics: the popularity of "vortex theory," whose purpose was to explain the variety of atoms by knots of different type.[32] Knot theory is an interesting area of mathematics that today indeed has applications in physics, but atomic structure isn't one of them. Nevertheless, vortex theory, at its peak, collected about twenty-five scientists, mostly in Great Britain but also in the United States, who wrote several dozen papers in the period from 1870 to 1890. Back then this was quite a sizable and productive community.

The followers of vortex theory were convinced by the theory's beauty despite the utter lack of evidence. In 1883, in a brief review for the magazine *Nature*, Oliver Lodge referred to vortex theory as "beautiful" and "a theory about which one may almost dare to say that it deserves to be true."[33] Albert Michelson (who would go on to win a Nobel Prize) wrote in 1903 that vortex theory "ought to be true even if it is not."[34] Another fan was James Clerk Maxwell, who opined:

> *But the greatest recommendation of [vortex] theory, from a philosophical point of view, is that its success in explaining phenomena does not depend on the ingenuity with which its contrivers "save appearances," by introducing first one hypothetical force and then another. When the vortex atom is once set in motion, all its properties are absolutely fixed and determined by the laws of motion of the primitive fluid, which are fully expressed in the fundamental equations.... The difficulties of this method are enormous, but the glory of surmounting them would be unique.*[35]

Regardless of what it ought to have been, vortex theory became obsolete with measurements of the atomic substructure and the advent of quantum mechanics.

And not only does the history of science thrive with beautiful ideas that turned out to be wrong, but on the flipside we have the ugly ideas that turned out to be correct.

Maxwell himself, for example, didn't like electrodynamics the way he conceived it because he couldn't come up with an underlying mechanical model. Back then, the standard of beauty was a mechanical clockwork universe, but in Maxwell's theory electromagnetic fields just *are*—they are not made of anything else, no gears or notches, no fluids or valves. Maxwell was unhappy about his own theory because he thought that only "when a physical phenomenon can be completely described as a change in the configuration and motion of a material system, the dynamical explanation of that phenomenon is said to be complete." Maxwell tried for many years to explain electric and magnetic fields by something that would fit the mechanistic worldview. Alas, in vain.

Mechanism was the fad of the time. William Thomson (later Lord Kelvin) thought that only when physicists have a mechanical model can they really claim to understand a particular subject.[36] Ludwig Boltzmann, according to his student Paul Ehrenfest, "obviously derived intense aesthetic pleasure from letting his imagination play over a confusion of interrelated motions, forces and reactions until the point was reached where they could actually be grasped."[37] Later generations of physicists simply noted that such underlying mechanistic explanations were superfluous, and they became accustomed to working with fields.

Half a century later, quantum electrodynamics—the quantized version of Maxwell's electrodynamics—also suffered from a perceived lack of aesthetic appeal. The theory gave rise to infinities that had to be removed by provisional methods introduced for no reason other than to give useful results. It was a pragmatic approach that Dirac didn't like at all: "Recent work by Lamb, Schwinger, Feynman and others has been very successful...but the resulting theory is an ugly and incomplete one and cannot be considered as a satisfactory solution of the problem of the electron."[38] When asked for his opinion about the recent developments in quantum electrodynamics, Dirac said, "I might have thought that the new ideas were correct if they had not been so ugly."[39]

In the following decades, better ways were found to deal with the infinities. Quantum electrodynamics, it turned out, is a well-behaved theory in which the infinites can be unambiguously removed by introducing two parameters that have to be determined experimentally: the mass and charge of electrons. These methods of "renormalization" are still used today. And despite Dirac's disapproval, quantum electrodynamics is still part of the foundations of physics.

To wrap up my historical excursion: aesthetic criteria work until they don't. The most telling evidence for the ineffectiveness of experience-based aesthetic guidance may be that no theoretical physicist has won a Nobel Prize twice.[40]

Why Trust a Theorist?

It's December and it's Munich. I am at the Center for Mathematical Philosophy to attend a conference that promises to answer the question "Why trust a theory?" The meeting is organized by the Austrian philosopher Richard Dawid, whose recent book *String Theory and the Scientific Method* caused some upset among physicists.[41]

String theory is currently the most popular idea for a unified theory of the interactions. It posits that the universe and all its content is made of small vibrating strings that may be closed back on themselves or have loose ends, may stretch or curl up, may split or merge. And that explains everything: matter, space-time, and, yes, you too. At least that's the idea. String theory has to date no experimental evidence speaking for it. Historian Helge Kragh, also at the meeting, has compared it to vortex theory.[42]

Richard Dawid, in his book, used string theory as an example for the use of "non-empirical theory assessment." By this he means that to select a good theory, its ability to describe observation isn't the only criterion. He claims that certain criteria that are not based on observations are also philosophically sound, and he concludes that the scientific method must be amended so that hypotheses can be evaluated on purely theoretical grounds. Richard's examples for this non-empirical evaluation—arguments commonly made by string theorists in favor of their theory—are (1) the absence of alternative explanations, (2) the use of mathematics that has worked before, and (3) the discovery of unexpected connections.

Richard isn't so much saying that these criteria *should* be used as simply pointing out that they *are* being used, and he provides a justification for them. The philosopher's support has been welcomed by string theorists. By others, less so.

In response to Richard's proposed change of the scientific method, cosmologists Joe Silk and George Ellis warned of "breaking with centuries of philosophical tradition of defining scientific knowledge as empirical" and, in a widely read comment published in *Nature*, expressed their fear that "theoretical physics risks becoming a no-man's-land between mathematics, physics and philosophy that does not truly meet the requirements of any."[43]

I can top these fears. If we accept a new philosophy that promotes selecting theories based on something other than facts, why stop at physics? I envision a future in which climate scientists choose models according to criteria some philosopher dreamed up. The thought makes me sweat.

But the main reason I am attending this conference is that I want answers to the questions that attracted me to physics. I want to know how the universe began, whether time consists of single moments, and if indeed everything can be explained with math. I don't expect philosophers to answer these questions. But maybe they are right and the reason we're not making progress is that our non-empirical theory assessment sucks.

The philosophers are certainly right that we use criteria other than observational adequacy to formulate theories. That science operates by generating and subsequently testing hypotheses is only part of the story. Testing all possible hypotheses is simply infeasible; hence most of the scientific enterprise today—from academic degrees to peer review to guidelines for scientific conduct—is dedicated to identifying good hypotheses to begin with. Community standards differ vastly from one field to the next and each field employs its own quality filters, but we all use some. In our practice, if not in our philosophy, theory assessment to preselect hypotheses has long been part of the scientific method. It doesn't relieve us from experimental test, but it's an operational necessity to even get to experimental test.

In the foundations of physics, therefore, we have always chosen theories on grounds other than experimental test. We have to, because often our aim is not to explain existing data but to develop theories that we hope will later be tested—if we can convince someone to do it. But how are we supposed to decide what theory to work on before it's been tested? And how are experimentalists to decide which theory is worth testing? Of course we use non-empirical assessment. It's just that, in contrast to Richard, I don't think the criteria we use are very philosophical. Rather, they're mostly social and aesthetic. And I doubt they are self-correcting.

Arguments from beauty have failed us in the past, and I worry I am witnessing another failure right now.

"So what?" you may say. "Hasn't it always worked out in the end?" It has. But leaving aside that we could be further along had scientists not been distracted by beauty, physics has changed—and keeps on changing. In the past, we muddled through because data forced theoretical physicists to revise ill-conceived aesthetic ideals. But increasingly we first need theories to decide which experiments are most likely to reveal new phenomena, experiments that then take decades and billions of dollars to carry out. Data don't come to us anymore—we have to know where to get them, and we can't afford to search everywhere. Hence, the more difficult new experiments become, the more care theorists must take to not sleepwalk into a dead end while caught up in a beautiful dream. New demands require new methods. But which methods?

I hope the philosophers have a plan.

∞∞

THE VENUE of the conference is the main building of the Ludwig Maximilian University in Munich, originally completed in 1840, rebuilt after being partly destroyed in the Second World War. The ceiling has round arches and the floors are marble; the corridors are flanked by pillars, decorated with the occasional stained-glass window and fire extinguisher. In the conference room, dead men stare down from gold-framed oil paintings. The meeting starts at exactly 10:00 a.m.

Also at the workshop is Gordon "Gordy" Kane, an American particle physicist. Gordy is author of several popular science books about particle physics and supersymmetry, and he is also known for his efforts to connect string theory with the standard model. He claims he can derive from string theory the conclusion that supersymmetric particles must appear at the LHC.

During Kane's talk, an argument erupts among the physicists. A few of them debate Kane until a philosopher loudly complains he wants to hear the rest of the talk. "This is part of what we call the scientific method," snarls David Gross, a longtime supporter of string theory who "heartedly recommend[s]" Richard Dawid's book, but then he sits back down.[44] Doubts remain about whether Kane's predictions indeed follow from string theory or whether he has used additional hand-selected assumptions in order to reproduce what we already know about the standard model.

Gordy might be overstating the rigor of his derivation, but he is doing the hard work, being one of the few who are trying to find a path from the beautiful idea of string theory back to the messy reality of particle physics. Gordy's path leads through supersymmetry, which is a necessary part of string theory. While finding superpartners would not prove string theory right, it would be a first milestone on the path to connect string theory with the standard model.

In his 2001 book, Gordy described supersymmetry as "wonderful, beautiful, and unique," and back then he showed himself confident that the LHC would discover superpartners. His confidence was based on an argument from naturalness. If one assumes that the supersymmetric theory contains only pretty numbers—neither very large nor very small—one can estimate the masses of the superpartners. "Luckily, the expected masses are small enough that they imply the superpartners should be detected soon," Gordy wrote in that book, and he explained that "the superpartner masses cannot be very much larger than the Z-boson mass if this whole approach is valid." This means if superpartners exist, the LHC should have long since seen them.

∞∞

GORDY'S ESTIMATE relies on one of the main motivations for supersymmetry: it avoids the need to fine-tune the mass of the Higgs boson, one of the twenty-five particles of the standard model. This argument is representative of many similar arguments that we will encounter later, and therefore we will give it a closer look.

The Higgs is the only known particle of its type, and it suffers from a peculiar mathematical problem that the other elementary particles are immune to: quantum fluctuations make a huge contribution to the Higgs's mass. Contributions like this are normally small, but for the Higgs they lead to a mass much larger than what is observed—too large, indeed, by a factor of 10^{14}. Not a little bit off, but dramatically, inadmissibly wrong.*

That the math gives a wrong result for the Higgs boson's mass is easy to remedy. One can amend the theory by subtracting a term so that the remaining difference gives the mass we observe. This amendment is possible because neither of the terms is separately measurable; only their difference is. Doing this, however, requires that the subtracted term be delicately chosen to almost, but not exactly, cancel the contribution from the quantum fluctuations.

For this delicate cancellation, one needs a number identical to that generated by the quantum fluctuations for fourteen digits, and then different in the fifteenth digit. But that such a close pair of numbers would come about by chance seems highly unlikely. Imagine you make two draws from a huge bowl that contains lottery tickets with every possible fifteen-digit number. If you draw two numbers that are identical except for the last digit, you would think there must be an explanation—maybe the tickets weren't mixed very well, or maybe someone played a trick on you.

Physicists feel the same way about the suspiciously small remainder between two large numbers that is necessary to give the correct Higgs mass—it seems to require an explanation. But when it comes to the laws of nature we do not draw numbers from a bowl. We have

---

* Quick reminder: a 10 with a raised *x* is a 1 followed by *x* zeroes. So, for example, $10^2 = 100$.

only these laws and we have no way of telling how likely or unlikely they are. That the Higgs mass requires explanation, therefore, is indeed feeling, not fact.

Physicists call a number that seems to require explanation "fine-tuned," while a theory that has no fine-tuned numbers is "natural."* A natural theory is also often described as one that uses only numbers close to 1. The two notions of naturalness are the same because if two numbers are close together, then their difference is much smaller than 1.

In summary, numbers that are very large, very small, or very close together are not natural. In the standard model, the Higgs mass is not natural, which makes it ugly.

Supersymmetry much improves the situation because it prevents the overly large contributions from quantum fluctuations to the Higgs mass. It does so by enforcing the required delicate cancellation of large contributions, without the need to fine-tune. Instead there are only more moderate contributions from the masses of the superpartners. Assuming all masses are natural then implies that the first superpartners should appear at energies not too far away from the Higgs itself. That's because if the superpartners are much heavier than the Higgs, their contributions must be canceled by a fine-tuned term to give a smaller Higgs mass. And while that is possible, it seems absurd to fine-tune susy, since one of the main motivations for susy is that it avoids fine-tuning.

In case I lost you on the quantum math, the argument goes like this: we don't like small numbers, so we invented a way to do without them, and if that's right, then we should see new particles. This isn't a prediction, it's a wish. And yet these arguments have become so common that particle physicists use them without hesitation.

That heavy superpartners would bring back the naturalness problems was the main reason many physicists believed the new particles had to show up at the LHC. "If SUSY exists, many of its most important motivations demand some SUSY particles at the TeV

* "Fine-tuning" has a slightly different meaning in cosmology. We will come to this later.

range"—that's an example from a 2005 lecture series by Carlos Wagner at the Enrico Fermi Institute in Chicago that I quote merely to represent what I heard at dozens of seminars.[45] "Theorists love SUSY for her elegance," wrote Leon Lederman shortly before the LHC became operational. "The LHC will allow us to establish whether SUSY exists or not: even if 'squarks' and 'gluinos' [two types of superpartners] are as heavy as 2.5 TeV, the LHC will find them."[46]

The Higgs boson was found with a mass of approximately 125 GeV. But no superpartners have shown up, nor has anything else that the standard model can't explain. This means we know now that if the superpartners exist, their masses must be fine-tuned. Naturalness, it seems, is just not correct.

That the naturalness arguments turned out to be wrong has left particle physicists at a loss for how to proceed. Their most trusted guide appears to have failed them.

ooOo

THE LHC began its first run in 2008. Now it's December 2015 and no signs of supersymmetry have been found, in blatant contradiction to Gordy's 2001 prediction. But after a two-year upgrade the LHC began running again early this year, this time at an energy of 13 TeV, and we haven't yet seen results from this second run, so there is still hope.

During a conference break I find time to ask Gordy his opinion of the present situation. "What did you think when superpartners didn't show up at the LHC?"

"There was no reason for them to show up in run one," Gordy says. "Not a single motivation except for this naive naturalness argument. But once you actually ask for a theory that makes predictions, you look at string theory. They shouldn't show up in run one. They might show up in run two."

It is the ability to adapt to new evidence that marks the true scientist.

"What if they're not in run two either?"

"Then this model is wrong. I'm not sure. I would have to think hard what is wrong. Since the predictions of the model are so generic, I do expect superpartners to show up. [If they didn't] I would certainly take the data seriously and wonder what one could change in the model. I don't know anything. But I'd then want to spend some time looking and see if something came up that could be changed."

After the first days at the Munich workshop it has become clear to me that nobody here has practical advice for how to move on. Maybe I was expecting too much of the philosophers.

What I learn, however, is that Karl Popper's idea that scientific theories must be falsifiable has long been an outdated philosophy. I am glad to hear this, as it's a philosophy that nobody in science ever could have used, other than as a rhetorical device. It is rarely possible to actually falsify an idea, since ideas can always be modified or extended to match incoming evidence. Rather than falsifying theories, therefore, we "implausify" them: a continuously adapted theory becomes increasingly difficult and arcane—not to say ugly—and eventually practitioners lose interest. How much it takes to implausify an idea, however, depends on one's tolerance for repeatedly making a theory fit conflicting evidence.

I ask Gordy: "Do you think the elegance of a theory is something that theorists do pay attention to—and that they *should* pay attention to?"

"They do pay attention. I did," he says. "Because it feels good to do the work and it keeps it exciting." He briefly pauses. "I hesitate to say 'should.' Tentatively I *should* do what makes me feel good about doing it. But the 'should' is more about feeling good than a logical principle. If there's a better way, somebody else should do it. But I doubt there's a better way, and this one feels good."

IN BRIEF

- Scientists have used beauty as a guide for a long time. It hasn't always been a good guide.
- In theoretical physics, symmetries have been very useful. They are presently considered beautiful.
- Particle physicists also think a theory is beautiful if it contains only "natural" numbers—numbers that are close to 1. An unnatural number is called "fine-tuned."
- If we are starved of data and need a theory to decide where to look for new data, mistakes in theory development can lead to a dead end.
- Some philosophers are proposing to weaken the scientific method so that scientists can select theories by criteria other than a theory's ability to describe observation.
- The questions of how to move on despite lack of data and whether to amend the scientific method are relevant beyond the foundations of physics.

3
The State of the Union

In which I sum up ten years of education in twenty pages and chat about the glory days of particle physics.

The World According to Physicists

The most astounding fact about high-energy physics is that you can get away with knowing nothing about it.

Take calcium, one of the substances in your bones. The calcium atom consists of 20 neutrons and 20 protons bound together in the atomic core—also called the "nucleus"—surrounded by 20 electrons. The electrons are fermions and form discrete shells around the core. It's the structure of these shells that determines the chemical properties of calcium.

The protons and neutrons in the nucleus, however, don't sit still: they are constantly in motion, shifting and bumping into each other, emitting and absorbing force-carrying particles which hold them together. There is no quiet in the subatomic world, ever. Yet, despite the constant action, calcium atoms all behave alike. And lucky for you, because otherwise your bones might fall apart.

Intuitively you have known this all your life, that whatever the neutrons do inside atoms can't be all that important or else you'd have heard of it. But conceptually this lack of influence is absolutely astonishing. Given the enormous number of individual constituents, why doesn't all this atomic substructure lead to behavior that's exceedingly

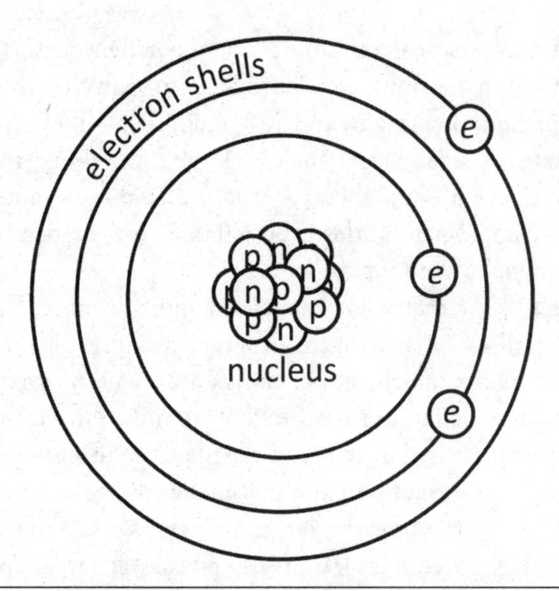

FIGURE 2. The atomic shell model, in which electrons (*e*) sit on separate shells around the atomic nucleus, which is made of protons (p) and neutrons (n). It is an example of the separation of scales. What the particles in the nucleus do doesn't affect the electron shells and the atom's chemical properties.

difficult to pin down? Why are atoms all so similar? The many particles they are made of each do their own thing, but atoms follow remarkably simple laws—so simple that atoms can be neatly classified in the periodic table based solely on the electrons' shell structure.

Nature, it seems, caters extremely kindly to our desire to understand. Whatever happens in the nucleus stays in the nucleus; we only see the net effect. Some atoms bind with hydrogen, others don't—but what exactly goes on in the nucleus has nothing to do with this bond. Some atoms form regular periodic lattices, others don't—and what happens in the nucleus has no influence on the lattice structure.

It's not only for atoms that we can ignore exactly what the constituents do. Properties of composite particles such as neutrons and protons are also almost unaffected by the motion of their constituents, the quarks and gluons. And when we describe, for example, the way that atoms push around grains of pollen on the surface of water (Brownian motion), it is sufficient to think of the atoms as particles in their own right and just ignore that they are made of smaller things. At even larger scales it's the same thing: the orbits of planets don't depend on planetary structure, and—zooming out some more—on cosmological scales even galaxies can be dealt with as if they were particles without constituents.

This isn't to say that what happens at short distances doesn't have any effect at all on what happens at larger distances; it's just that the details don't matter much. Large things are made of smaller things, and the laws for the larger things follow from the laws for the smaller things. The surprise is that the laws for the large things are so simple.

Much of the information from the smaller things, it turns out, isn't relevant to understanding the larger things. We say that the short-distance physics "decouples" from the physics at larger distances or that "the scales separate." This separation of scales is the reason why you can go through life without knowing a thing about quarks or the Higgs boson, or—to the dismay of physics professors all over the world—without having any clue what quantum field theory is.

This separation of scales has far-reaching consequences. It means that we can devise approximate laws of nature that to good accuracy describe a system at some given resolution, and then revise these laws as we increase the resolution. The approximate laws, which are good only at a certain resolution, are called "effective laws."

As one decreases the resolution, it is then often practical to adapt the objects that the theory deals with and also the properties that we assign to them. In the theory at lower resolution, it might make more sense to collect many small constituents into one larger object, give a name to that large object, and assign it properties. This way, we can speak of atoms and their shell structure, of molecules and their vibrational modes, and of metals and their conductivity—even though in the underlying theory there aren't any such things as atoms, metals, or their conductivity, only elementary particles.

Each level of resolution therefore has its own language, which is the formulation most useful on that level. We call these resolution-dependent objects and their properties "emergent." The process that relates the theory for short distances to the one for larger distances is also referred to as "coarse-graining" (Figure 3).

"Emergent" is the opposite of "fundamental," which means that an object cannot be further decomposed and its properties cannot be derived from a more accurate theory. Being fundamental is a matter of current knowledge. What is fundamental today might no longer be fundamental tomorrow. What is emergent, however, will remain emergent.

Matter is made of molecules, which are made of atoms, which are made of standard-model particles. The standard-model particles plus space and time are, for all we currently know, fundamental—they are not made of anything else. In the foundations of physics, we are trying to find out if there is something even more fundamental.

As a physicist, I am often accused of reductionism, as if that were an optional position to hold. But this isn't philosophy—it's a property of nature, revealed by experiments. We have extracted these layers of

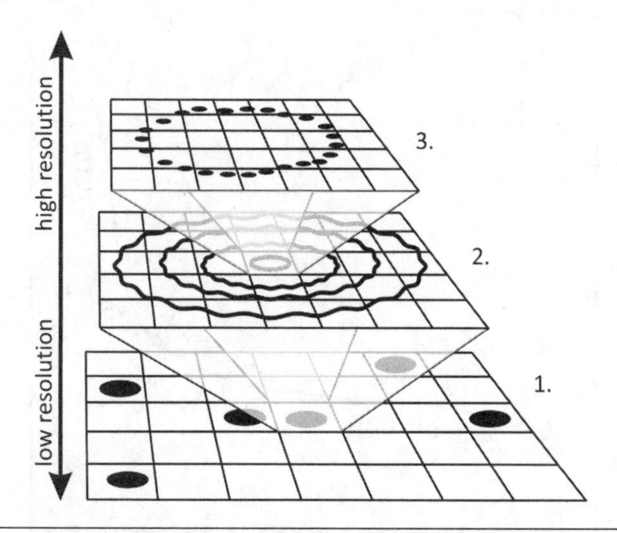

FIGURE 3. Illustration of coarse-graining. The low-resolution objects and their laws (level 1) can be described by the medium- or high-resolution objects and laws (level 2 or 3), but not the other way round. Lower-resolution levels emerge from higher-resolution levels.

resolution and their laws from countless observations and found that they describe our world dramatically well. Effective field theory tells us we can, in principle, derive the theory for large scales from the theory for small scales, but not the other way round.

That the history of science has slowly revealed this hierarchical structure is why today many physicists think that there must be one fundamental theory from which everything else derives—a "theory of everything." It's an understandable hope. If you'd been sucking away on a giant jawbreaker for a century, wouldn't you hope to finally get close to the gum?

Everything Flows

Effective laws that depend on resolution offer another way to see why naturalness is pretty. For this, physicists assign every theory a place in

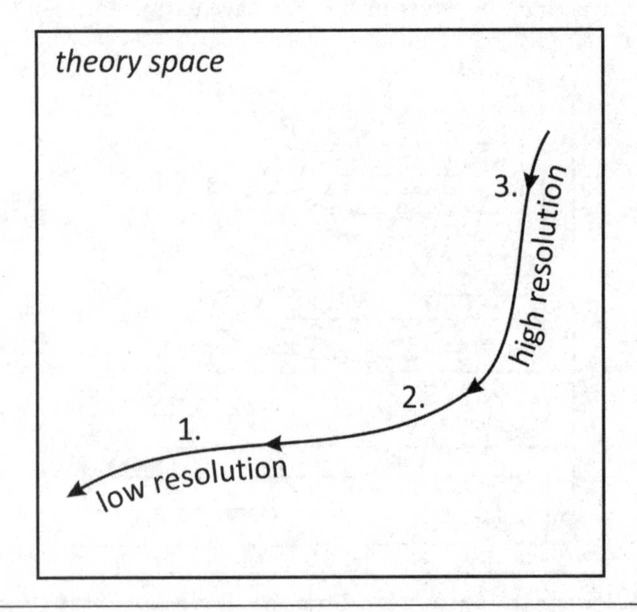

FIGURE 4. Every point in theory space is a different theory. If we change the resolution, we trace out a path. Numbers refer to the levels of Figure 3.

an abstract "theory space," which is useful to picture relations between different theories. Since theories depend on resolution, each one traces out a curve in theory space if the resolution changes (Figure 4). The curves of all theories together are referred to as the "flow" of theories.

In theory space, naturalness means that a theory at low resolution should not sensitively depend on the theory at high resolution (which is presumed to be more fundamental). The idea is that whatever we choose for the parameters of the more fundamental theory at high resolution, at low resolution the physics should resemble the one we observe. This is the original motivation for naturalness: our choice shouldn't matter.

The flow in theory space makes it possible to quantify just how much a theory at low resolution depends on the choice of parameters at high resolution; this is the way that Giudice's measures work.* The idea is illustrated in Figure 5. In this figure, low resolution is what we can probe now—this is where we have the standard model. It might seem odd to call this "low resolution," given that it's the highest resolution we have ever reached! But it *is* low compared to the resolution we think is necessary to reveal a theory of everything, believed to be far beyond even the LHC's reach.

The standard model (at low resolution) is natural or not fine-tuned if it doesn't matter much exactly where in theory space we start at high resolution. In this case, the flow will always bring us to a place close by (i.e., within measurement precision of) the standard model (Figure 5, left). If, on the other hand, we have to precisely pick the theory at high resolution to end up close by the standard model, then we have to fine-tune the starting point. In this case, the standard model is unnatural (Figure 5, right).

In the fine-tuned case, the starting points of theories that reproduce the standard model (i.e., are consistent with observations) must be close together. This small distance corresponds to the ugly small numbers that we discussed previously, such as the mass of the Higgs boson.

---

* There are several different measures and some disagreement about which is the best, but this won't matter for the following.

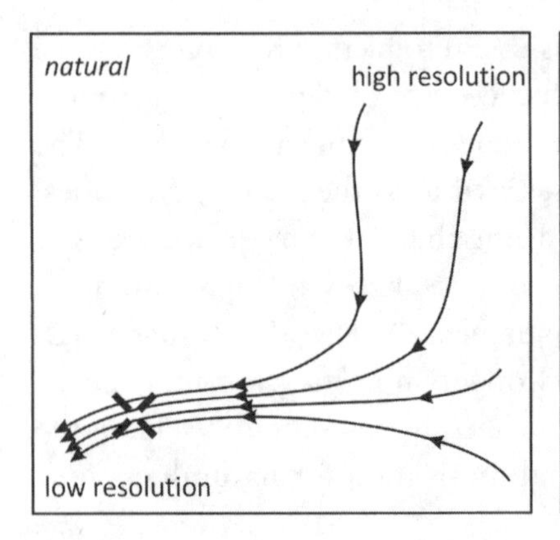

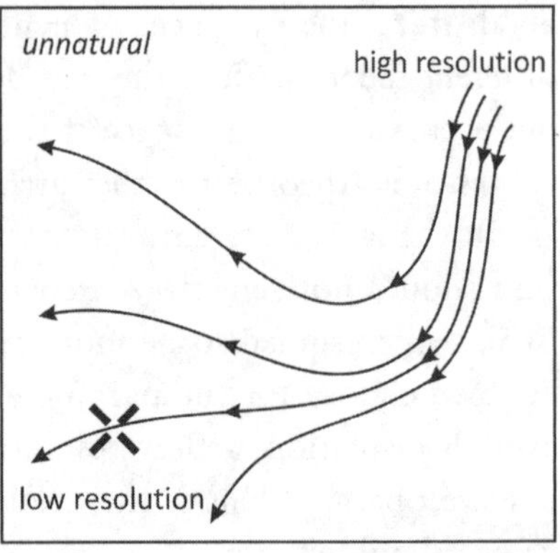

FIGURE 5. Illustration of the flow in theory space for the case when the theory (e.g., the standard model, marked with X) at low resolution is natural/not fine-tuned (left) and when it is unnatural/fine-tuned (right).

In the next section I will summarize the laws of space, time, and matter that we have discovered so far, and the type of experiments that revealed them. If you are familiar with the standard model and the concordance model, you may want to skip to the end of the section.

The Tools of the Trade

In 1858, the Irish American writer Fitz-James O'Brien imagined the perfect microscope. In his short story "The Diamond Lens," Linley, the mad microscopist, consults the spirit of Antonie van Leeuwenhoek who, two hundred years earlier, had discovered bacteria while improving the very first microscopes.[1] Throughout his life, van Leeuwenhoek had remained famously secretive about his methods to manufacture lenses. But with the help of a medium named Madame Vulpes, Linley learns from the deceased van Leeuwenhoek that it takes "a diamond of one hundred and forty carats, submitted to electro-magnetic currents for a long period," to construct a microscope "whose magnifying power should be limited only by the resolvability of the object."

Lacking sufficient research funding, Linley murders a friend and steals the required diamond. Then he gazes into a drop of water:

> *I can not, I dare not, attempt to inventory the charms of this divine revelation of perfect beauty. Those eyes of mystic violet, dewy and serene, evade my words. Her long, lustrous hair following her glorious head in a golden wake, like the track sown in heaven by a falling star, seems to quench my most burning phrases with its splendors.*

It remains to be seen whether nature at shortest distances is as beautiful as O'Brien envisioned it, but we already know his perfect microscope will remain fiction. The resolving power of lenses depends on the messenger they rely on: light. Long wavelengths aren't sensitive to small distances, much the way heavy boots aren't sensitive to the grooves on escalator treads. The resolution of microscopes is limited by the wavelength of the light in use, and to probe shorter distances we need shorter wavelengths.

Visible light has wavelengths of about 400 to 700 nanometers.* That is about 10,000 times larger than a hydrogen atom. So visible light works well if we want to study cells, but it's insufficient if we want to study atoms. We can reach better resolution by using light with shorter wavelengths, like X-rays, which improve over visible light by a factor of 100 to 10,000. But light with even shorter wavelengths becomes increasingly difficult to direct and to focus.

To further improve resolution, therefore, we draw on the central lesson of quantum mechanics: there aren't really waves and there aren't really particles. Instead, everything in the universe (including, for all we known, the universe itself) is described by a wave function that has properties of both particles and waves. Sometimes this wave function appears more like a wave, sometimes more like a particle. But fundamentally it is neither—it's a new category in its own right.

---

* 1 nanometer is 10^{-9} m. That's a billionth of a meter.

Strictly speaking, therefore, we shouldn't refer to "elementary particles" at all, which is why one of my profs suggested we instead call them "elementary things." But it's an expression nobody uses, and I don't want to torture you with it either. Just keep in mind that whenever physicists refer to particles, they actually mean a mathematical object called the wave function, which is neither a particle nor a wave but has properties of both.

The wave function itself does not correspond to an observable quantity, but from its absolute value we can calculate probabilities for the measurement of physical observables. That's the best we can do in quantum theory—except in special circumstances, the outcome of a single measurement cannot be predicted.

Quantum theory helps us improve the resolution of microscopes because it reveals that the heavier a particle (thing?) is and the faster it moves, the smaller its wavelength. Therefore, electron microscopes, which use beams of electrons rather than light, reach much higher resolutions than light microscopes. If the electrons move at even moderate speed, using electric and magnetic fields, such microscopes can resolve structures as small as an atom. In principle, we can increase this resolution arbitrarily by further speeding up the electrons. This is the basic reason why modern physics drives and is driven by the construction of larger and larger particle accelerators: more collision energy means probing shorter distances.

Unlike light microscopes, which use mirrors and lenses, particle accelerators use electric and magnetic fields to accelerate and focus beams with electrically charged particles. But as one speeds up the particles used to study an object, it becomes increasingly difficult to extract information from the measurement. That is because the particles meant to test the probe start to noticeably alter the probe. Visible light shining on a slice of onion doesn't do much to the onion, other than maybe heating it up ever so slightly. But a speedy beam of electrons shot at a target plate will, at sufficiently high energy, destroy the target. The information about what happened at very short distances, then, is in the debris. And that is pretty much what high-energy physics is: trying to extract information from the debris of collisions.*

The resolution that can be achieved with accelerators is inversely proportional to the total energy of the colliding particles. A good benchmark to remember is that an energy of 1 GeV (that's 10^9 eV or 10^{-3} TeV, about the mass of the proton) corresponds to a resolved distance of approximately 1 femtometer (10^{-15} m, about the size of a proton). An order up in energy means an order down in distance and vice versa. The LHC was designed to reach a peak collision energy of about 10 TeV. This corresponds to about 10^{-19} m and is the shortest distance at which we have tested the laws of nature—so far.

The task of theoretical physicists is to find the equations that accurately describe the outcomes of particle collisions. When a calculation matches the data, we gain confidence in the theory. When theoretical physicists understand particle collisions better, experimentalists can design detectors more efficiently. And when experimentalists understand accelerator technology better, theorists get better data.

This strategy has been tremendously successful and resulted in the standard model of particle physics, our best current knowledge about the elementary building blocks of matter.

The Standard Model

The standard model is based on a principle called "gauge symmetry." According to this principle, every particle has a direction in an internal space, much like the needle on a compass, except that the needle doesn't point anywhere you can look.

"What the heck is an internal space?" you ask. Good question. The best answer I have is "useful." It's what we invented to quantify

* At the highest energies, physicists don't shoot particles at targets but just collide two particle beams. This delivers cleaner signals and also increases the total collision energy.

the observed behavior of particles, a mathematical tool that helps us make predictions.

"Yes, but is it real?" you want to know. Uh-oh. Depends on whom you ask. Some of my colleagues indeed believe that the math of our theories, like those internal spaces, is real. Personally, I prefer to merely say it describes reality, leaving open whether or not the math itself is real. How math connects to reality is a mystery that plagued philosophers long before there were scientists, and we aren't any wiser today. But luckily we can use the math without solving the mystery.

So, each particle has a direction in this useful internal space. What we call a gauge symmetry then demands that the laws of nature do not depend on the labels we use to mark that space—like we could turn a compass so that the needle points northwest rather than north. With such a change, a north particle could morph into a combination of other particles, becoming, say, a northwest particle. Indeed, this is what happens with an electron: a transformation in its internal space can make the electron morph into a mixture of an electron and a neutrino. But if this transformation is a symmetry, then the particle mixup shouldn't change the physics. The symmetry requirement therefore limits the possible laws we can write down. The logic is similar to coloring a mandala. If you want the color fill to respect the symmetry of the design, you have fewer options than when you ignore the symmetry.

For the laws of nature, the symmetry requirement is not easy to fulfill. A big complication is that turns of the internal space could differ from one moment to the next and from one place to another, and that too should not affect the laws which the particles obey. If we express this symmetry requirement in mathematical form, we see that it entirely fixes how the particles must behave. The interaction between the particles that obey the symmetry must be mediated by another particle whose properties are determined by the type of symmetry involved. This additional particle is called the gauge boson of the symmetry.

The previous paragraph condenses a math-heavy derivation, and with such brevity you will get only a rough impression of how it works. But the upshot is that if we want to construct a theory that maintains (is gauged under) a certain symmetry, then this necessarily gives rise to a specific type of interaction between the particles that respect the symmetry. Moreover, the symmetry requirement automatically adds the force-carrying gauge bosons that must come with it. It is this type of gauge symmetry that underlies the standard model.

Remarkably, the standard model works almost entirely with such symmetry principles. It combines the electromagnetic force with the strong nuclear force (responsible for holding atomic nuclei together against electric repulsion) and the weak nuclear force (responsible for nuclear decay). For these three interactions there are three gauge symmetries, and all the particles are classified by the way the symmetries act on them. (I told you we care more about the ideas than the

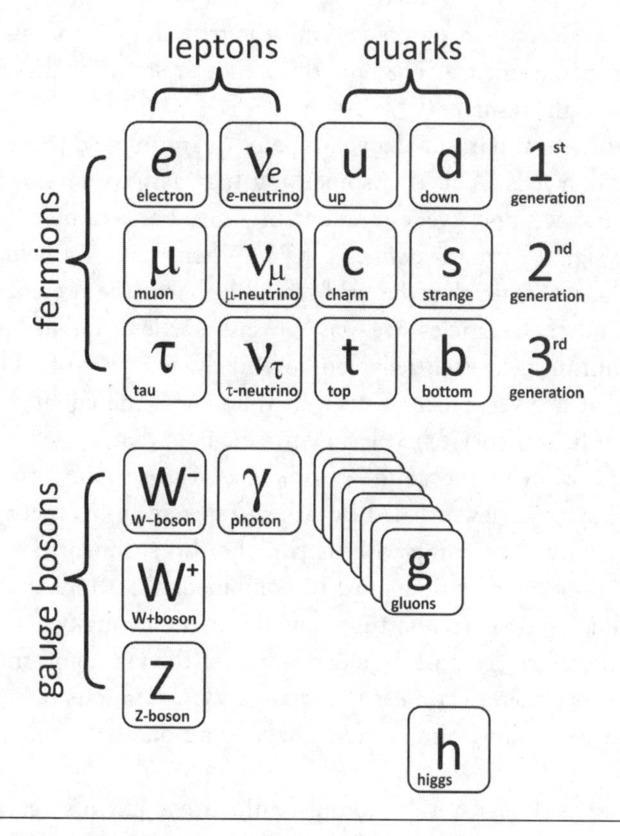

FIGURE 6. The standard model of particle physics.

particles, but since it's the particles that we measure, you will find a quick rundown in Appendix A and a summarizing table in Figure 6.)

The standard model is an exquisite construct of abstract math, a quantum field theory with gauge symmetries. I used to think saying this makes me sound educated. But I have noticed that incomprehensibility tends to attract suspicion. How can we be so sure that everything is made of merely twenty-five particles if we can't see most of them?

The answer is fairly simple. We use all this math to compute the outcome of experiments, and these calculations correctly describe observations. That's how we know the theory works. Indeed, that's what we mean by "the theory works." Yes, it's abstract, but that we merely see detector readouts and not the particles themselves is an irrelevant inconvenience. The only thing that's relevant is that the math gives the right result.

That the standard model is a type of quantum field theory sounds scarier than it is. A field is something that assigns a value to every point in space and every moment in time. For example, your cell phone's signal strength defines a field. When we call a field a quantum field, we mean that the field really describes the presence of particles, and the particles are—as we saw earlier—quantum things. The quantum field itself tells you how likely you are to find a certain particle at a given place at a given time. And the equations of the quantum field theory tell you how to calculate that.

In addition to the gauge symmetries, the standard model also uses the symmetries revealed by Albert Einstein in his theory of special relativity. According to Einstein, the three dimensions of space and the dimension of time must be combined into a four-dimensional space-time, and space and time must be treated equally. The laws of nature therefore (1) must be independent of the place and time where you measure them, (2) must not change with rotations in space, and (3) must not change under generalized rotations between space and time.

A space-time rotation sounds unhealthy, but it's really just a change of velocity. This is why popular science books about special relativity are often full of rocket ships and satellites passing each other. But all of this is really unnecessary decoration. Special relativity follows from the three symmetries listed above, without twins in spaceships and laser clocks and all that. It also follows that all observers agree on the speed of massless particles (such as photons, the carriers of light) and that nothing can exceed this speed. In other words, it follows that nothing travels faster than light.*

The gauge symmetries and the symmetries of special relativity dictate most of the standard model's structure, but it has some features that we have not (yet?) been able to explain by symmetries. One such feature, for example, is that fermions come in three generations, which are sets of similar particles with increasingly higher masses (the first generation contains the lightest fermions, the third the heaviest; the second is in between). Another unexplained feature is that each type of fermion comes in two versions that are mirror images of each other, referred to as "left-handed" and "right-handed," respectively—except for the neutrinos, whose right-handed versions no one has ever seen. But we will speak more about what's amiss with the standard model in Chapter 4.

The development of the standard model began in the 1960s and was largely completed by the late 1970s. Besides the fermions and gauge bosons, there is only one more particle in the standard model: the Higgs boson, which gives masses to the other elementary particles.[2] The standard model works without the Higgs; it just doesn't describe reality, because then all the particles are massless. Sheldon Glashow therefore once charmingly referred to the Higgs as the "flush toilet" of the standard model—it was invented for a purpose, not because it's pretty.[3]

The Higgs boson, proposed independently by several researchers in the early 1960s, was the last fundamental particle to be

* For the purposes of this book, that's all we need to know about special relativity, but that's certainly not all there is to say. For those interested in further reading, I suggest Chad Orzel's *How to Teach Relativity to Your Dog* (Basic Books, 2012) and Leonard Susskind and Art Friedman's *Special Relativity and Classical Field Theory: The Theoretical Minimum* (Basic Books, 2017).

discovered (in 2012), but it was not the last particle to be predicted. Last predicted—in 1973—were the top and bottom quarks, whose existence was experimentally confirmed in 1995 and 1977, respectively. In the late 1990s, neutrino masses—whose theory goes back to the 1950s—were added after experiments confirmed them. But since 1973 there hasn't been any successful new prediction that would supersede the standard model.

ooOo

THE STANDARD model is currently our best answer to the question "What are we made of?" But it does not account for gravity. That's because particle physicists don't need to account for gravity when making predictions for accelerator experiments: the masses of individual elementary particles are tiny, and so is their gravitational pull. Gravity is the dominant force over long distances, but over the short distances probed by particle collisions it is negligibly, indeed unmeasurably, small. However, while all the other forces can—and will—combine to neutral, this is not the case for gravity. While for large objects all the other forces neutralize and become unnoticeable, the gravitational force adds up and becomes noticeable.

Gravity also stands apart from the other interactions because in our current theories it is the only (fundamental) force that does not have quantum properties; it is an unquantized—we say "classical"—force. We will see what problems this brings in Chapter 7, but first let me tell you what we know about gravity and how we came to know it.

While particle physicists built larger colliders to look at increasingly shorter distances, astronomers built larger telescopes to look at increasingly longer distances.[4] The first telescopes were developed alongside the first microscopes, but the instruments designs quickly specialized. And in this area too, theory and experiment developed alongside each other.

Since very little light reaches us from faraway stellar objects, astronomers constructed telescopes with larger apertures or larger mirrors to collect as much light as possible. This method soon reached its limits because the huge apparatuses became impossible to handle. But the game changed entirely in the middle of the nineteenth century with the development of photographic plates. Now astronomers could collect light over long exposure times. But since Earth moves, longer exposure times would result in smears of light unless the telescopes incorporated a mechanism to compensate for this, which again required knowing the motion. And so the more astronomers learned about the night sky, the better they became at observing it.

Today, astronomers no longer capture images on photographic plates but in charge-coupled devices—the electronic hearts of digital cameras. Modern telescopes are so sensitive they can detect single photons, and exposure times sometimes reach millions of seconds (more than a week).[5] And of course telescopes are still getting bigger: We now have powerful machines to move around huge mirrors, which themselves are adjusted by thousands of little motors to prevent deformations due to gravity and temperature changes. Supercomputers and extraordinarily accurate time measurements made it possible for telescopes at distant locations to act in concert, which effectively creates even larger telescopes. To deal with atmospheric fluctuations that blur images, astronomers now use "adaptive optics," computer code that readjusts telescopes in response to atmospheric effects. Or they avoid atmospheric distortions completely by mounting telescopes on satellites and shooting them into outer space.

We have extended our reach from visible light to the long wavelengths in the infrared, micrometer, and radio spectrums, and to the short wavelengths in the X-ray and gamma ray range. And light isn't the only messenger we now use to study the cosmos. Other particles, including neutrinos, electrons, and protons, also tell stories about the sources they originated from and what happened on their way to Earth. Astronomers' most recent achievement has been the first direct detection of gravitational waves, which are disturbances in spacetime itself. These carry information about the often violent events they were created in, such as black hole mergers.

With these methods combined, astronomers have been able to look back to the time when the universe was only 300,000 years young, and out to distances of some 10 billion light-years. The data are utterly different from those seen in collider physics. But for us theorists, the task is the same: explain the measurements.

The Concordance Model

Our best current explanation for the astronomers' measurements is what we call the "cosmological concordance model."[6] This model uses the mathematics of general relativity, according to which we live in three space dimensions and one time dimension, and moreover, this space-time is curved.

I know, it's hard to imagine a curved four-dimensional space-time—it's not just you. Luckily, for many purposes two-dimensional surfaces are good analogies. Special relativity treats space-time like a flat sheet of paper. But in general relativity, in contrast, space-time can have hills and valleys.

To continue the analogy, if you have a map of a mountainous landscape that doesn't show altitudes, winding roads won't make much sense. But if you know there are mountains, you understand why the roads curve like that—in such terrain, it's the best they can do. That we cannot see the curvature of space-time is like having a map without altitude lines. If you could see space-time curvature, you would understand it makes perfect sense for planets to orbit around the Sun. It's the best they can do.

General relativity is based on the same symmetries as special relativity. The difference is that in general relativity space-time becomes responsive; it reacts to energy and matter by curving. In return, the motion of energy and matter depends on the curvature.

But the curvature doesn't only change from one place to another; it also changes with time. Most important, therefore, general relativity taught us that the universe isn't eternally unchanging. It expands in reaction to matter, and with this expansion matter becomes more thinly distributed.

That the universe expands means that matter must have been squeezed together in the past. The early universe, therefore, was filled with a very dense but almost homogeneous particle soup. The soup was also hot, which means the average energy of individual particle collisions was high. This creates a problem, because if the temperature exceeds roughly 10^{17} Kelvin, then the average collision energy exceeds the energy presently tested at the LHC.[7] For higher temperatures—that is, for a younger universe—we have no reliable knowledge about the behavior of matter. We have speculations, of course, and we'll talk about some of them in Chapters 5 and 9. But for now, let us focus on what occurs below this temperature, where the concordance model explains what happens.

General relativity gives us the equations that relate the expansion of the universe to the types of energy and matter in it. Cosmologists therefore can find out what's in the universe by trying different combinations of matter and energy and seeing which one works best to explain observations (or rather, they let a computer try). They do this every time new observations are made. And boy, have they found some surprises!

The most shocking finding is that the presently dominant source of gravity in the universe isn't anything we have previously encountered. It is an unknown type of energy—dubbed "dark energy"—and it constitutes a whopping 68.3 percent of the total matter-energy budget. We don't know if dark energy has a microscopic substructure; we only know the effect it has. Dark energy speeds up the expansion of the universe. That's why we need so much of it, because data show that the universe's expansion has been speeding up. However, dark energy is also very thinly distributed, so we cannot measure it in our immediate vicinity. It is only over large distances that we notice its net effect through the acceleration of expansion.

The simplest type of dark energy is the cosmological constant, which doesn't have any substructure and is constant both in space and in time. A cosmological constant is what the concordance model uses as dark energy, but it could be something more complicated.

The remaining 31.7 percent of what fills the universe is matter, but—more surprise—most of it isn't the kind of matter that we're familiar with. Instead, 85 percent of matter (26.8 percent of the total energy-matter budget) is what's called "dark matter." The only thing we know about dark matter is that it interacts rarely, both with itself and with other matter. In particular, it doesn't interact with light, hence the name. Some supersymmetric particles behave like dark matter should, but we don't yet know if that's the right explanation.

The remaining 15 percent of the matter in the universe (4.9 percent of the total energy-matter budget) is stable standard-model particles—stuff similar to what we are made of (Figure 7).

Once we know what types of energy and matter fill the universe, we can reconstruct the past. In the early universe, dark energy (in the form of a cosmological constant) was negligibly small compared to matter. That's because the density of matter decreases as the universe expands, whereas a cosmological constant remains, well, constant.

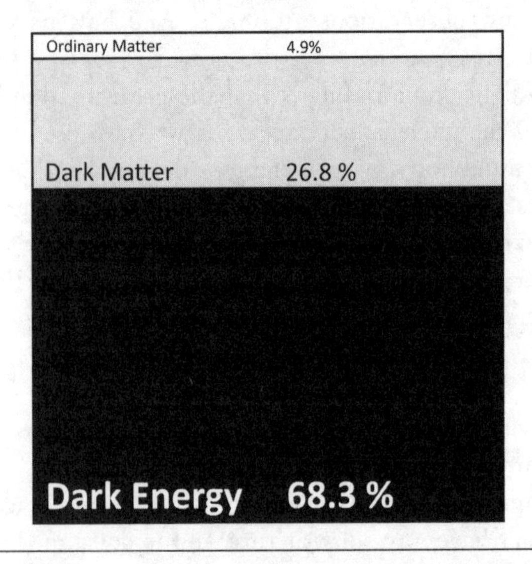

FIGURE 7. The energy content of the universe for people who don't like pie charts.

So if they are comparably large today, with a ratio of roughly 2:1, then in the early universe the density of matter must have been much larger than the energy density of the cosmological constant.

At 10^{17} Kelvin, therefore, we start with a soup composed almost entirely of matter and dark matter. Space-time reacts to this matter by beginning to expand. This cools the soup and enables the formation first of atomic nuclei and then of light atoms. Initially, the particle soup is so dense that light gets stuck in it. But once atoms form, light can travel almost undisturbed.

Dark matter, since it doesn't interact with light, cools faster than normal matter. In the early universe, therefore, dark matter is the first to start clumping under its own gravitational pull. Indeed, without the dark matter's early clumping, galaxies wouldn't form in the way that we observe, for the gravitational pull of the already clumpy dark matter is needed to speed up the clumping of normal matter. And it's only when enough normal matter has come together that the formation of large atomic nuclei in stellar cores can begin.

Under the pull of gravity, over the course of billions of years, galaxies assemble, solar systems form, stars ignite. Up until this point the universe has been expanding, though the expansion had begun to slow down. But around the time when galaxies are fully formed, dark energy takes over and the expansion rate of the universe begins to speed up. It is this period that we presently live in. From here on into the future, matter only dilutes more. Therefore, if dark energy is a cosmological constant, it will continue to dominate, and the expansion of the universe will continue to speed up—forever.

The wavelength of the first light that escaped the particle soup in the early universe has stretched with the universe's expansion but the light is still around today. Its wavelength is now a few millimeters, far outside the visible range and instead in the microwave-range. This cosmic microwave background (CMB) is measurable, and it's cosmologists' most precious source of information.

The mean temperature of the CMB is about 2.7 Kelvin, not far above absolute zero. But around the mean temperature are tiny deviations, about 0.003 percent of the absolute temperature. These deviations come from spots that were a little hotter or colder than average in the early universe. The fluctuations in the CMB temperature therefore encode fluctuations in the hot soup, which is what seeded galaxies.

Armed with that knowledge, we can use the cosmic microwave background to infer the history of the universe, as I have described above. Other data come from the observed distribution of galaxies, various measurements of the universe's expansion, the abundance of chemical elements, and gravitational lensing, just to mention the most important ones.[8]

The concordance model is also referred to as ΛCDM, where Λ (the Greek letter capital lambda) stands for the cosmological constant and CDM stands for "cold dark matter." The standard model and the concordance model together presently constitute the foundations of physics.[9]

It's Going to Be Hard Now

I used to attend the international conference series "Supersymmetry and Unification of Fundamental Interactions." Since 1993, it has taken place annually, and at its peak it gathered more than five hundred participants. Every year the talks laid out the benefits of supersymmetry: naturalness, unification, and dark matter candidates. Every year the searches for superpartners came back with negative results. Every year the models were updated to accommodate the lack of evidence.

The failure to date of the LHC to provide evidence of superpartners has taken a toll on theorists' mood. "It is not time to be desperate yet...but maybe it is time for depression already," remarked the Italian physicist Guido Altarelli in 2011.[10] Ben Allanach, from the University of Cambridge, has described his reaction to a 2015 analysis of LHC data as "a bit depressing for a supersymmetry theorist like me."[11] Jonathan Ellis, a theorist at CERN, has referred to the possibility that the LHC would find nothing but the Higgs boson as "the real five-star disaster."[12] The name that has stuck, however, is "the nightmare scenario."[13] We're now living this nightmare.

I haven't attended the annual conference since 2006—too depressing. But from these conferences I know Keith Olive and his work on supersymmetry. Keith is professor of physics at the University of Minnesota and director of the William I. Fine Theoretical Physics Institute. I call Keith to inquire what he makes of the no-show of susy at the LHC.

"We got the data in little bits," Keith remembers. "The limits just got stronger and stronger. Every few months when we got a new data analysis it got a little bit worse. It's certainly true that we expected susy at lower energy. It's a big problem. There's something in me that tells me that supersymmetry should be part of nature, though, as you say, there's no evidence for it. Should it be at low energy? I think nobody knows. We thought it would be."

Keith is from the generation before mine, the generation that witnessed the successes of symmetry and unification in the development of the standard model. But I have no such experience, no reason to think that beauty is a good guide. This voice telling Keith what is and isn't part of nature? I don't trust it.

"Why should susy be part of nature?"

"It's the strength of its symmetry," Keith says. "I think it's still very compelling. Whether it's accessible at low energies, I think it's still possible. If the Higgs had turned out to be 115, 120 GeV and susy wasn't found, this would have been much more troubling. That the value of the Higgs mass is close to the upper limits gives some hope. Things actually have to be heavy, and it makes some sense that the LHC wouldn't see it."

The Higgs boson decays quickly after being produced, and its presence must be inferred from the decay products that reach the detector. But how the Higgs decays depends on its mass. A heavy Higgs, as long as it is being produced at all, would have made for a signal that was easier to find. Hence, even before the LHC began its search, the mass of the Higgs was constrained both from below and from above.

The LHC eventually confirmed the Higgs with a mass of 125 GeV, just on the upper edge of the range that had so far not been excluded.

A heavier Higgs allows heavier superpartners, so as far as susy is concerned, the heavier the Higgs, the better. But the fact that no superpartner has yet been found means that such a superpartner would have to be so heavy that the measured Higgs mass could be achieved only by fine-tuning the parameters of the supersymmetric models.

"So now we know there is some fine-tuning," Keith says. "And that itself becomes a very subjective issue. How bad is the fine-tuning?"

ooo

AS WE have seen, physicists don't like numerical coincidences that require very large numbers. And, since the inverse of a very large number is a very small number and hence one can be converted into the other, they don't like very small numbers either. Generally, they don't like numbers much different from 1.

But it is only the numbers without units that physicists worry about—numbers that are also called "dimensionless," as opposed to "dimensionful" numbers, which do have units. This is because the value of a dimensionful number is fundamentally meaningless—it depends on the choice of units. Indeed, with suitable units, any dimensionful number can be made to be 1. The speed of light, for example, is equal to 1 in units of light-years per year. So when physicists fret about numbers, it's only the ones without units that are worrisome, such as the ratio of the mass of the Higgs boson to the mass of the electron, which comes out to be approximately 250,000:1.

The conundrum with the mass of the Higgs boson, which we discussed earlier, isn't that the mass itself is small, since such a statement depends on the units involved and is therefore meaningless. The Higgs mass is 1.25×10^{11} eV, which looks large, but that's the same as 2.22×10^{-21} gram, which looks tiny. No, what is small isn't the Higgs mass itself, but the ratio of the Higgs mass over the (mass associated with the) energy that comes from the quantum corrections to that mass. I hope you excuse my earlier sloppiness.

The origin of arguments from naturalness is that physicists would like all dimensionless numbers to be close to 1. But the numbers don't have to be exactly 1, and so one can debate just how large a number is still acceptable. Indeed, in many equations we already have dimensionless numbers, and these can bring in factors that aren't necessarily close to 1. For example, 2π to a power that depends on the number of spatial dimensions can swiftly wind up as more than 100 (especially if you have more than three dimensions of space). And if you make your model a little more complicated, maybe you can get a number that's even higher.

So, how much fine-tuning is too much depends on your tolerance for combining factors into larger factors. Therefore, the assessment of just how much trouble supersymmetry is in, now that the LHC results require it to be fine-tuned in order to get the Higgs mass right, is subjective. We may be able to exactly calculate how much fine-tuning is required. But we can't calculate how much fine-tuning theorists are willing to accept.

ooOo

"ONE OF the standard motivations for supersymmetry is always that it avoids fine-tuning," Keith says. "We like to think that if there is some theory beyond the standard model and if you write down [quantum] corrections, you'd not like having to fine-tune them to this accuracy."

"What is wrong with fine-tuning?"

"It seems not to be attractive!" he says, and laughs. "Naturalness is sort of a guiding principle. If that's what you call attractive, that's the definition of attractive: it attracts us. That's where we flock to.

"At the end of the day," Keith continues, "the only thing we know to be true is the standard model. Which is frustrating to everybody. [There] should be something beyond the standard model, if it's just to explain dark matter or to explain [why the universe contains more matter than antimatter]. Something really should be there. For a lot of people it's just hard to imagine that it's just some random other thing, something that is entirely disconnected. The thing to do in my mind is to add symmetry, or to add unification."

I ask Keith which experimental strategy to pursue, but he has no advice.

"All the easy stuff has been done," he says. "It's going to be hard now. It's going to be hard. In the 1950s, when particle physics started, it was a lot easier. It wasn't that hard to build a few-GeV machine and do the collisions. And there was just stuff coming out all over the place—none of that was known physics. You have all this strange stuff—that's why they call them 'strange' particles! The number of particles discovered per year was phenomenal. And that led to all sorts of theoretical progress. Now... it's tough without any experimental guidance. Then we're doing things based on what we think is pretty."

IN BRIEF

- Experiment and theory normally advance together.
- What we presently believe to be the most fundamental laws of nature build on symmetry principles.
- If new data become sparse, theoretical physicists rely on their perception of beauty to assess theories.
- Beauty is not a scientific criterion but may be an experience-based one.

4
Cracks in the Foundations

In which I meet with Nima Arkani-Hamed and do my best to accept that nature isn't natural, everything we learn is awesome, and that nobody gives a fuck what I think.

A Great Job, If You Can Get It

A group of schoolchildren is taking photos of the Niels Bohr Institute when my taxi arrives. On the front of the house, letters form the institute's name and the year of construction, 1920. It was here in Copenhagen that physicists convened almost a century ago to develop the foundations of atomic physics and quantum mechanics, the theories that underlie all modern electronics. Every microchip, every LED, every digital camera, and every laser—we owe them all to the equations conceived here, when Heisenberg and Schrödinger came to discuss physics with Bohr. It's a good place to photograph while your teacher is watching.

As I stand in front of the closed doors, the winter rain blowing into my face, I realize the building might date back to 1920, but the electronic locks don't. It takes some wandering around to find the registration desk in an adjacent building. A young Danish woman informs me that I am not on her list of expected visitors, and she asks about the purpose of my visit.

"I'm here to visit Nima Arkani-Hamed," I offer, struck by the awkwardness of jumping onto a plane just to hold a recording device under someone's mouth. But Nima himself is visiting at the institute—I don't know whom—and he isn't on her list either.

I am only half lying when I say I come from Nordita, the Niels Bohr Institute's former sister institute, which relocated to Stockholm in 2007. Though my contract just ran out, I still smile from the website. She hands me a keycard.

I'm too early, and so I seek out the library down the hallway. Familiar books greet me. The wooden floor squeaks, and I stop moving, not wanting to interrupt potentially world-changing thoughts. It smells like science, by which I mean coffee. I recall a story that during World War II the building was equipped with explosives on the reasoning it'd be better to blow it up than leave it to the Nazis; the rumor is that nobody is sure all the explosives were removed after the war. I move on carefully.

I have just resolved to track down the coffee machine when Nima arrives. Since I first came across his papers in the late 1990s, Nima's career has been nothing but stellar. In 1999, at the age of twenty-seven, he became a faculty member in the physics department at Berkeley. He moved on to Harvard in 2002, has been at Princeton since 2008, and was elected a fellow of the American Academy of Arts and Sciences in 2009. He's won loads of awards, including the inaugural 2012 Breakthrough Prize for "original approaches to outstanding problems in particle physics." The problems are still outstanding. So is Nima.

He shows me to the office he's using during his stay here at the institute. I plop down on a couch, unsure as to exactly what I'm supposed to do next. Pushing the record button on my recorder seems a good idea. And as if he'd only been waiting for his cue, Nima starts talking, hands waving, hair flying.

The question of beauty and naturalness, he begins, has been much on his mind in light of the recent results from the LHC.

"The general topic of naturalness and beauty is badly misrepresented," Nima says. "This conflation of beauty between arts and science probably helps to sell books." And he isn't happy with that. "If you're a hard-nosed layperson and your knowledge about physics was based on Brian Greene's book *The Elegant Universe*—not to pick on Brian—but you could get away with the idea that physicists are just making shit up. This is unfortunate because it's far away from reality, the reality of a decent, honest physicist.

"Yes," he says, "you can easily get the impression that sometimes the experiments are practically impossible. And even if they are practically possible, they might take so long that, practically speaking, you could go most of your life without having to confront [the results of an] experiment. And before that, anything goes. You can make up all kinds of la-la stories, and maybe every now and then, once every fifty years or so, experiment will come along and chop things down. So, great job if you can get it, right? You can just sit around, make shit up, and you're never checked. That's the impression that I would get."

After my appointment at Nordita ended, I left Stockholm and moved to Germany. But the new research grant hasn't come through yet, and so I'm currently unemployed. It's not the first time this has happened. For fifteen years now, I've been hopping from one short-term contract to another, from one country to another, driven by the conviction that physics is my best shot at making sense of the world. It's not so much a profession as an obsession. My situation is the norm, Nima's the exception. Most of us don't get accused of having a great job.

Oblivious to my thoughts, Nima goes on: "So there's no experiment and you just sit around and talk about beauty and elegance and mathematical loveliness. And it sounds like sociological bullshit. I think this impression is just completely wrong—but completely wrong for an interesting reason. That reason makes us in high-energy physics different from most other parts of science.

"It's true," he explains, "that in most other parts of science what's needed to check whether ideas are right or wrong are new experiments. But our field is so mature that we get incredible constraints already by old experiments. The constraints are so strong that they rule out pretty much everything you can try. If you are an honest physicist, 99.99 percent of your ideas, even good ideas, are going to be ruled out, not by new experiments but already by inconsistency with old experiments. This is what really makes the field very different, and what gives us an internal notion of right and wrong before new experiments. So, quite contrary to the sense that this skeptical layperson would get, the idea that you can just make up crap is wrong. It's an incredible struggle."

Tell me about struggle, I think, and nod.

Making Problems

The standard model, despite its success, doesn't get much love from physicists. Michio Kaku calls it "ugly and contrived," Stephen Hawking says it's "ugly and ad hoc," Matt Strassler disparages it as "ugly and baroque," Brian Greene complains that the standard model is "too flexible," and Paul Davies thinks it "has the air of unfinished business" because "the tentative way in which it bundles together the electroweak and strong forces" is an "ugly feature."[1] I yet have to find someone who actually likes the standard model.

What makes the standard model so ugly? Its worst offense is the many parameters—numbers for which there is no deeper explanation—and many of these parameters aren't anywhere close to 1. We already discussed the headache that is the Higgs mass. But there are more such annoying numbers, beginning with the masses of the other elementary particles, or the ratios of these masses to the Higgs mass, respectively (keeping in mind that we fret only about dimensionless numbers). This ratio takes on values such as 0.000004008 for the electron or approximately 1.384 for the top quark. Nobody can explain why these mass ratios are what they are.

But the mass ratios also don't seem entirely random, and this makes physicists believe there is an underlying explanation. All three neutrinos, for example, are very light, with the sum of their masses less than 10^{-11} times that of the Higgs. The fermion generations have masses that differ roughly by a factor of ten. And there is Koide's strange formula, a relation between the masses of the electron, muon, and tau particles.[2] If you divide the sum of their masses by the square of the sum of the square roots of their masses, the result is 2/3 up to the fifth digit after the decimal point. Why? Similar numerological relations have been found to hold for other particles, though with lower levels of accuracy. They make us think that we're missing a deeper explanation.

Besides the masses, there are the mixing matrices. When traveling from one place to another, some particles can morph—or "oscillate"—into different particles. The probabilities of this happening are collected in what are known as the mixing matrices.[3] Here too, the numbers in the matrices are so far unexplained, but they don't seem entirely random. Some particles mix into others almost evenly, but others don't mix much even though they could. Why is that so? We don't know.

The next problem is that the standard model has too much symmetry! The symmetry in question is known as CP symmetry. Its symmetry transformation is a combination of changing the electric charge of a particle to its opposite (denoted C, for "charge") and changing the particle to its mirror image (denoted P, for "parity"). If one does this transformation, then the equations of the weak nuclear force change—that is, the electroweak force does not obey the symmetry. Quantum electrodynamics cannot violate this symmetry. The strong force can violate it, but for unknown reasons it doesn't. If it did, then this should affect, for example, the distribution of electric charge in the neutron, and this hasn't been seen.

The strength of this CP violation of the strong force is measured by what is known as the theta parameter (θ). According to current data, this parameter is offensively small, far smaller than 1.

A proposed mechanism to solve this so-called strong CP problem is to make the theta parameter dynamic and let it roll down a potential to the minimum, where it settles at a small value.[4] This solution would be natural because it does not require new large or small numbers. It was, however, noticed by Steven Weinberg and, independently, by Frank Wilczek that a dynamic theta must be accompanied by a particle, which Wilczek dubbed "axion" (the first and, hopefully, last particle to be named after a laundry detergent). The axion hasn't been found, however, and the strong CP problem has remained unsolved.

But when we look at the standard model, it's not only numbers that irk us. There are also the three unexplained generations of fermions and the three gauge symmetries. Wouldn't it be so much nicer if the electroweak and strong interactions could be merged, producing a grand unified theory or, even better, a supersymmetric grand unification? (More about this in Chapter 7.)

And of course we also have complaints about the cosmological concordance model. Here too we have all these unexplained numbers. Why is the amount of dark energy what it is? Why is there five times as much dark matter as normal matter? And what are dark matter and dark energy, anyway? In the concordance model we merely describe their bulk behavior, but their microscopic properties do not play a role. Do they have microscopic properties at all? Are they made of something? And if so, what? (We will discuss this in Chapter 9.)

Then there are the problems with combining the concordance model and the standard model. The gravitational force between elementary particles is extremely weak compared to that of the other interactions. The ratio of the electric and gravitational forces between an electron and proton, for example, is about 10^{-40}. Yet another unexplained, small number, this one known as the "hierarchy problem."

Worse still, general relativity refuses to consistently combine with the standard model, which is why physicists have tried for eighty years to develop a quantized version of gravity—a theory of "quantum gravity." Ideally, they would like to also merge quantum gravity with all the other interactions to make a theory of everything. (We will return to this in Chapter 8.)

Finally, even if we'd solved all of these issues, we'd still be complaining about quantum mechanics (the topic of Chapter 6).

ꝏꝏ

THESE PROBLEMS have been known for at least twenty years, and none of them presently seems close to resolution. Partly responsible for the lack of progress is the increasing difficulty of conceiving (and financing) new experiments—the easy ones have all been done. You can expect such slowdown as a research area matures.

As we saw, however, theorists have no shortage of puzzles even without new experiments. Indeed, most of my colleagues believe these problems to be solvable on purely theoretical grounds. They just haven't yet managed to do that. Theoretical progress has thus slowed, and for much the same reasons that new experiments are hard to come by: the easy things have been done.

Every time we solve a problem, it becomes more difficult to change anything about today's theories without reproblematizing issues that we've already resolved. And so the fundamental laws of nature we now have seem unavoidable consequences of past achievements. This inescapability of the existing theories is often referred to as "rigidity." It raises our hopes that we already know everything necessary to find a more fundamental theory—and that being smart will be sufficient to find it.

One way to look at this situation is to say that rigidity is desirable because it signals that a theory is close to uniquely fitting our observations. The other way to look at it is that rigidity means we have reached a dead end, must revisit long-solved problems and look for the path not taken.

ʘʘʘ

"TO SAY it more simply," Nima tells me, rephrasing his earlier point, "that we have both relativity and quantum mechanics is an incredible constraint. I think this point is not widely understood: that both relativity and quantum mechanics shockingly—shockingly!—constrain what you can do. Rigidity and inevitability is by far the most important thing. You can call it whatever you want, but for me it's a stand-in for beauty."

"But why do we have symmetries to begin with? Why quantum fields? Why a curved space-time?" I ask, listing some commonly made mathematical assumptions.

We use these and other abstractions because they work, because we have found they describe nature. From a purely mathematical standpoint they are certainly not inevitable; if they were, we could derive them by logic alone. But we can never prove any math to be a true description of nature, for the only provable truths are about mathematical structures themselves, not about the relation of these structures to reality. Hence rigidity is a meaningful criterion only once we fix a backbone of assumptions from which to make deductions.

Gravity, for example, is almost unavoidable once you buy into the idea that we inhabit a curved space-time. But that doesn't tell you why we live in a curved space-time to begin with; it's just something we've found to work. And we only know it works for the cases we have tested.

"You're right," Nima says. "Any discussion of rigidity has to be within the context of something that's accepted to be true. Not because we know it is—of course we don't know." He launches an impromptu lecture about the origins of string theory, and then vanishes to get coffee.

I find it hard to disagree with him. Quantizing gravity is arguably a technically very hard problem. The symmetries of special relativity are extremely difficult to maintain in a quantum theory of gravity, and that difficulty makes one suspect that if we find one way to do it, then maybe that's the only way.

Then again, this might say more about humans than about physics.

"Why does susy continue to draw so much attention?" I ask Nima when he comes back.

Sipping his coffee, he says: "If supersymmetry is there not far from the [energies tested at the LHC], what's interesting about this is that it immediately puts incredibly tight constraints on what comes next. If there's a fourth generation [of fermions], it doesn't tell me anything. So there are some developments that are intellectual dead ends."

"Is it good," I wonder, "that theorists prefer not to look at what might be such an intellectual dead end? What is wrong with other ideas besides that they don't like them?"

"Who gives a fuck what you like or don't like?" Nima says. "Nature doesn't care, and we all agree about this. The reason that supersymmetry was so popular was not just sociology. What was decisive was that [with supersymmetry] you could solve problems that you couldn't solve otherwise. Whether or not nature cares about these problems is another question. Without susy you have these naturalness issues. Such an issue has come up three times or so before, and every time it's come up we've found a solution."

Where the Numbers Have No Name

Today we call a theory natural if it does not contain numbers that are either very large or very small. Any theory that contains unnatural numbers is believed to not be fundamental—it is a crack in the floor that deserves digging.

The idea that a law of nature should be natural in this fashion has a long history. It started as an aesthetic criterion, but now it has become mathematically formalized as "technical naturalness." And by promoting an aesthetic criterion to a mathematical prescription, the nonscientific origin of naturalness has largely been forgotten.

The first appeal to naturalness might have been the rejection of the heliocentric (Sun-in-center) model on the ground that the stars appear fixed. If Earth goes around the Sun, then the stars' apparent positions should change over the course of the year. The magnitude of this change, known as "parallax," depends on the distance to the stars; the farther away the star, the smaller the apparent change in position. You can see a similar effect when you are on a train watching the landscape go by: nearby trees move much faster through your field of view than the skyline of a distant city.

Back then astronomers thought the stars were fixed on a celestial sphere that contained the whole universe. In this case, if we aren't in the center of the sphere, the stars' relative positions should change during the year because sometimes we'd be closer to one side of the sphere than the other. Astronomers didn't see these changes, and so they concluded that Earth is at the center of the universe.

The stars do indeed slightly change their position over the course of the year, but the shift is so minuscule that astronomers couldn't measure it until the nineteenth century. The best they could do before that was to estimate that the absence of observable parallax meant that either Earth itself wasn't moving during the year or the stars had to be very, very far away—much farther away than the Sun and the other planets, in which case the parallax would be tiny. This would have allowed the Sun to be in the center, but it was an option they discarded because it required them to accept unexplainably large numbers.

In the sixteenth century, Nicolaus Copernicus made a compelling case for the heliocentric model on the ground that it simplified the motions of the planets, but the issue of the parallax still lingered. It wasn't only that the stars would have had to be much farther away than any other object in the solar system. An additional problem was that Copernicus and his contemporaries wrongly estimated the size of the stars.

Light from a distant source that passes through a circular aperture—such as that of the eye or a telescope—will smear out and appear with an increased width, but this wasn't understood until the nineteenth century. Due to this imaging artifact, the astronomers of Copernicus's time mistakenly assumed stars to be much larger than they really are. In the heliocentric model, then, the fixed stars had to be very distant and still appear large in the telescope, which meant that they had to be huge, much larger than our own Sun.

Tycho Brahe thought that such vastly different numbers were absurd and hence rejected the idea that Earth moves around the Sun. He instead proposed his own model, in which the Sun revolved around Earth but the other planets moved around the Sun. In 1602, he argued against heliocentrism on the ground that

> *it is necessary to preserve in these matters some decent proportion, lest things reach out to infinity and the just symmetry of creatures and visible things concerning size and distance be abandoned: it is necessary to preserve this symmetry because God, the author of the universe, loves appropriate order, not confusion and disorder.*[5]

This notion of "decent proportion" and "appropriate order" is basically today's criterion of naturalness.

We now know that most stars are comparable in size to our Sun, and there is nothing unnatural about the huge distance between us and the stars. The typical distances between our solar system and other stars in the Milky Way, as well as our distance to other galaxies, are determined by the way matter clumps under its own gravitational pull while the universe expands. The distances in question arise dynamically and are not fundamental parameters of any underlying theory.

But the idea that large numbers require explanation remained influential.[6] In 1937, Paul Dirac noticed that the age of the universe divided by the time it takes light to cross the radius of a hydrogen atom is approximately 6×10^{39}—roughly the same as the ratio of the electric force between a proton and an electron to the gravitational force between them, which is about 2.3×10^{39}. Not exactly the same, true, but close enough for Dirac to postulate that these numbers must have a common origin. And not only should these numbers be related, he argued, but "any two of the very large dimensionless numbers occurring in Nature are connected by a simple mathematical relation, in which the coefficients are of the order of magnitude unity."[7]

This has become known as Dirac's large number hypothesis.

But in his numerical game Dirac used a constant that isn't actually constant: the age of the universe. This means that to maintain the postulated equality, other constants of nature must change with time as well. The result is a plethora of consequences for the formation of structures in the universe that have rendered the hypothesis incompatible with observation.[8]

When applied to the particular numbers that he picked, Dirac's large number hypothesis is today believed not to be of fundamental relevance. However, the gist of his idea that large numbers need explanation, or at least it's preferable if several have a common origin, is still very much used. Indeed, physicists noted that the occurrence of conspicuously large or small numbers can indicate the presence of new, heretofore unaccounted-for effects. This added support to their belief that fine-tuning is a strong signal that an overhaul is necessary.

The logic of arguments from naturalness resembles the attempt to predict the plot of a long-running TV series. If the hero—here, naturalness—is in a pickle, certainly he will survive, so something must happen to turn around a situation that looks bleak.

In unquantized electrodynamics, for example, the mass of the electron is unnaturally small. That's because the electron carries an electric field and the field's energy should make a large (actually, infinite) contribution to the mass. Getting rid of this "self-energy" would require fine-tuning the math, and that's ugly. So here is our hero, naturalness, locked up in a burning house. If this calculation is correct, he will die.

But the calculation isn't correct because it neglects quantum effects. With these effects included, the electron is surrounded by pairs of virtual particles that are created and annihilated without becoming directly observable. They do, however, make indirect contributions that cure the self-energy troubles of the unquantized theory. The smallness of the electron mass is hence "natural" in quantum electrodynamics.[9] The hero jumps off the roof and lands in a dumpster, unhurt.

In particle physics in particular, the absence of numerical coincidences now has a mathematical formulation known as "technical naturalness."[10] Amazingly enough, the whole standard model is technically natural, except for the trouble with the Higgs mass. Even for composite particles bound by the strong nuclear force, all the masses are technically natural—with one exception: the masses of three mesons (the neutral pion and the two charged pions) are suspiciously close to each other; if you take the difference between the squares of the charged and neutral pions' masses and divide that by the square of the masses themselves, the result you get is unnaturally small. The hero is in trouble again, with his back to the wall and a gun at his head.

But here too it turns out that the calculation does not correctly predict what happens. What saves the day is that beyond a certain energy, new physics appears in the form of a particle—the rho meson—and with the rho meson comes a new symmetry that explains why the pion masses are so close together. The explanation is technically natural, no fine-tuning required. The gun misfires, and the hero escapes.

The quantum corrections to the electron self-energy and the rho meson, however, were not predictions; they were postdictions, or what you could call hindsight insights. (You watched reruns.) The only actual prediction based on naturalness was that of the charm quark, the fourth quark to be discovered. Its existence was proposed in 1970 to explain why the probability of certain particle interactions was unnaturally small.[11] With the charm quark included, these interactions simply were forbidden, and so their non-observation was explained naturally, without the need to fine-tune.

In summary, naturalness is respected by the standard model and scores a total of one prediction. On this basis, Nathan Seiberg, of the Institute for Advanced Study in Princeton, claims, "The notion of naturalness has been a guideline in physics over the past couple of centuries."[12] Consequently, the opposite of naturalness, fine-tuning, has become appalling. According to Lisa Randall, of Harvard University, "Fine-tuning is almost certainly a badge of shame reflecting our ignorance."[13] Or, as particle physicist Howard Baer told me, "Fine-tuning, I think, is just a sickness in theories that one has to look at, and that guides you to how these theories might be fixed up and what the right path is into the great unknown, into the frontier issues."[14]

The cosmological constant is not natural. But that has something to do with gravity, and therefore particle physicists don't feel responsible for it. Now that we know the Higgs mass isn't natural either, however, the problem is at their front door.

Nobody Promised a Rose Garden

"I'm not a defender of naturalness," Nima says. "Naturalness is not a principle, not a law. It's been thought of as a guide. Sometimes it was a good guide, sometimes it was a bad guide. You have to be open to the possibilities. Some people say naturalness is pure philosophy, but it's definitely not philosophy. It's done many things for us."

He goes through the examples that speak for naturalness, careful also to mention what speaks against it, and concludes: "Naturalness wasn't—shouldn't have been—the argument for the LHC. To the credit of CERN, this argument came from theorists. Having said that, it wasn't stupid to think that naturalness would be right. Because it had these successes."

Despite the successes of naturalness, Nima tells me that ten years ago he abandoned natural beauty in favor of a new idea called "split susy." Split susy is a variant of supersymmetry in which some of the expected susy partners are naturally so heavy that they are beyond the reach of the LHC. This explains why the susy partners have not yet been seen. But split susy then requires fine-tuning to get the observed mass of the Higgs right.

About the reaction of his colleagues to fine-tuning the theory, Nima recalls: "I have been literally yelled at by people at conferences. It's never happened to me before or since."

So that, I think, is what happens if you don't meet the beauty standard of the day.

"Has the LHC changed your perspective on naturalness?" I ask.

"It's interesting—there is this popular narrative now that theorists before the LHC were totally sure that susy will show up, but now there's a big blow. I think that the people who are professional model builders, the people I consider to be the best people in the field, they were worried already after LEP. But it turned into a program, how do you avoid [conflict with existing data]. They were not huge things, but they were little niggling things. The good people, they were not at all sure susy would show up at the LHC. And nothing has changed qualitatively since 2000 [when the LEP ended its last run]. Some loopholes have been closed, but nothing has changed qualitatively.

"You ask, why do people still work on it?" Nima continues. "It's in fact very funny. As I said, the best people had a pretty good idea what was going on—they were not sitting on their hands waiting for gluinos to pour out of the LHC.[15] They also had a pretty level reaction to the data."

But not one of those "best people" spoke up and called bullshit on the widely circulated story that the LHC had a good chance of seeing supersymmetry or dark matter particles. I'm not sure which I find worse, scientists who believe in arguments from beauty or scientists who deliberately mislead the public about prospects of costly experiments.

Nima continues: "The people who were sure it would be there are now positive it's not there. There are people now who speak about being depressed or worried or scared. It drives me nuts. It's ludicrously narcissistic. Who the fuck cares about you and your little life? Other than you yourself, of course."

He isn't speaking about me, but he might as well be, I think. Maybe I'm just here to find an excuse for leaving academia because I'm disillusioned, unable to stay motivated through all the null results. And what an amazing excuse I have come up with—blaming a scientific community for misusing the scientific method.

"Everything we are learning about nature is amazing." Nima's voice breaks into my thoughts. "If you have new particles, you have more clues. If you don't, you have still clues. It's a sign of the narcissism of our times that people use this language. In better times you wouldn't have been allowed to use this language in polite company. Who cares about your feelings? Who cares if you spent forty years of your life on this? Nobody promised a rose garden. This is a risky business. You want certainty, you do something else with your life. People spent centuries barking up the wrong tree. That's life."

We've been speaking for hours, but Nima's energy seems infinite, almost unnatural. Words are tumbling over each other because they can't get out of his mouth fast enough. He bounces on his chair, swivels around; sometimes he jumps up and scribbles on the blackboard. The more I watch him, the older I feel.

"It's especially frustrating if their perceptions of depression then taint what we should do next," he says. "I think it's ridiculous. It's very important to learn whether naturalness is wrong."

Particle physicists, needless to say, are already lobbying for a new collider. The Chinese circular collider that Nima favors would reach about 100 TeV collision energy, but it's not the only option currently under discussion.[16] Another well-received proposal is the International Linear Collider that the Japanese have expressed interest in building. And CERN has plans for a super-LHC with a circumference of 100 kilometers, reaching energies comparable to those of the Chinese collider. Maybe then we'll finally find susy.

"Susy really is a slowly boiling lobster story," Nima says, recalling the theory's past. "It should have been seen at LEP 1; it should have been seen in 1990. Many of the theorists in this business that I really respect expected Susy to show up at LEP. I expected it to show up at LEP.

"We didn't see it," he says. "But people thought of it as an opportunity, not a problem. They thought, 'Okay, it tells us that the theory also must have a special structure.'...In the 1990s, that was a perfectly reasonable possibility. But then the limits improved and you kept not seeing them. And in the 1990s, LEP measured the gauge couplings precisely enough so that we knew the gauge couplings would not meet in the [standard model], but they would meet with susy."

ꝏꝏ

IN A gauge theory, the most important parameter is the "coupling constant," which determines the strength of the interaction. The standard model has three of them, two for the electroweak interaction and one for the strong interaction. These constants are invariant over space and time, but their value depends on the resolution of the process by which they are measured. This is an example of the earlier-mentioned flow in theory space.

Since we need high energies to probe short distances, low resolution corresponds to low energy, and high resolution corresponds to high energy. In high-energy physics, maybe not surprisingly, it is

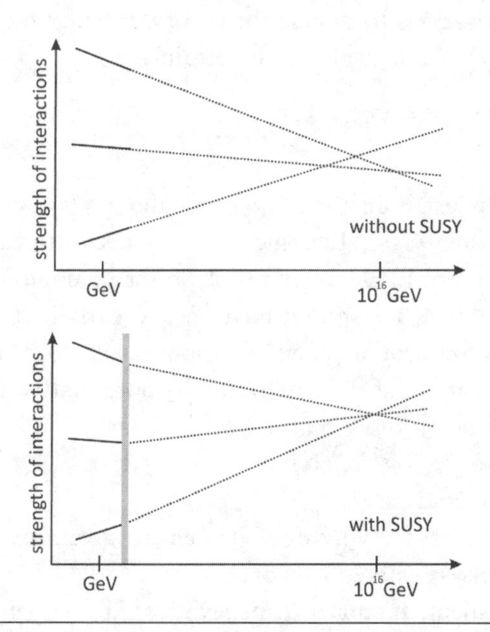

FIGURE 8. Schematic extrapolation (dotted line) of the strengths of the known interactions from the measured regime (solid line) to high energies. With supersymmetry, the curves meet in one point, assuming that susy particles start to show up at colliders soon (at energies near the gray bar). Past the unification energy, the forces' individual strengths lose meaning.

more common to think of the flow in theory space as a change in energy, not in resolution. The coupling constants then "run" with energy, as physicists say, and their running can be calculated.

If one does this calculation for the standard model, the curves agree with measurements at presently accessible energies. Extrapolated to higher energies, the strengths of the forces meet at three different points (see Figure 8, top). If one adds supersymmetry, however, they meet at one point (up to uncertainty from the measurement of the low-energy values), which is called "gauge-coupling unification" (Figure 8, bottom). If there is fundamentally only one gauge symmetry, there is fundamentally also only one coupling constant, and hence the three different couplings eventually must converge. That susy makes the constants match at high energies has been one of the strongest motivations to pursue the theory. Is gauge coupling unification necessary? No. Is it pretty? It certainly is.

ↀↀ

I THINK THE highest point of excitement about susy was probably in '91, '92," Nima says. "But since then it's been decreasing. When it didn't show up at LEP 2, many people in the community were saying there is a problem, we should have seen it earlier. If none of these superpartners are around, then why does gauge coupling unification work? And what about dark matter, why does that work?"

"Well, we don't know that it works," I say.

"Of course, of course," Nima agrees.

"It gives a candidate?"

"Okay," he says, "why does it seem to give a candidate? Why does it look like it wanted to work?"

I remain silent. It's hard to believe it's all a meaningless coincidence. Susy continues the quest for unification so naturally, works so nicely, fits in so snugly—it can't possibly all be wishful thinking driven by herd-thinking physicists. It's either me who's the idiot, or a thousand people with their prizes and awards. The odds aren't on my side.

"To come back to the question why do people still work on it," Nima says, breaking the silence, "to begin with, the interest in susy has plummeted. I mean, there are still people working on it. There are lots of diseases in academia and one of them is that you keep doing what you've been doing. Somehow it turns out that everybody does the analytic continuation of what they've been doing for their PhD.* But the good thing is that the experimentalists don't care."

"Well, you need the theorists to tell the experimentalists where to look."

"Right," he says.

But I'm happy to agree we're on the right track. All is good. Let's get back to business, build the next collider, find out what's up with that Higgs. We don't need philosophers to help us.

On the plane back to Frankfurt, bereft of Nima's enthusiasm, I understand why he has become so influential. In contrast to me, he believes in what he does.

The Diphoton Diarrhea Begins

It's December 15, 2015. Like thousands of my colleagues all over the world, I watch a live broadcast streamed from CERN. Today, collaborators from the two largest LHC experiments, CMS and ATLAS, will present their first results from the second run, measurements that have probed distances shorter than any experiment before.

Jim Olsen of the CMS team takes the stage. He begins to lay out the detector's design and the collaboration's analysis methods. I wish I could fast-forward. This could be the day the standard model begins to crumble—I want to see the data, not photos of magnets!

In the first run, some small deviations from the standard model's predictions had shown up. Deviations like this can—and frequently

---

* Nerd joke. A function defined over the real numbers can, by requiring certain niceness properties, be uniquely continued into the complex plane. This completion is referred to as the function's "analytic continuation." It means they keep doing what they were doing already.

do—appear by chance, the result of random fluctuations. Scientists hence assign a measure of reliability to any deviation from their current best theories, quantified by the probability that the deviation is merely coincidence. The data outliers from the first run still had a 1-in-100 or so probability of being due to chance. Fluctuations like this appear all the time and go away all the time, so they were nothing to get excited about.

Now Olsen goes through the CMS results. Indeed, all the fluctuations from the first run have decreased in significance, meaning they were almost certainly random noise. The collaboration has further analyzed the second-run data for signs of ideas popular in recent years. They haven't found anything: no evidence for extra dimensions, no superpartners, no mini black holes, no fourth generation of fermions. Delivered to much press attention, most of the announcement is a string of null results. I wonder what Gordon Kane makes of this.

But then, at the very end, Olsen reveals a new deviation from the standard model: too many events of a decay, which leaves behind two photons. Dubbed the "diphoton anomaly," this excess doesn't fit any of the existing predictions. It's incompatible with the standard model. It's incompatible with all theories we know. With that, Olsen hands over to Marumi Kado from the ATLAS collaboration.

Kado's summary is almost identical to Olsen's. The old fluctuations have faded, but ATLAS too has seen a diphoton excess where there shouldn't be one. That both experiments independently saw it substantially decreases the risk that the signal is mere coincidence. Taken together, they arrive at a chance of 3 in 10,000 that the excess is a random fluctuation. That's still far off the standard of certainty that particle physicists require for a discovery, which is roughly 1 in 3.5 million. But this could be it, I think, the first step on the way to a more fundamental law of nature. Immediately we start discussing what it could be.

A day later, the open-access server arXiv.org lists ten new papers on the diphoton anomaly.

IN BRIEF

- Theoretical physicists have a lot of complaints about the laws of nature they've found so far. They particularly dislike unnatural numbers.
- Naturalness has been used as a guide for theory development since at least the sixteenth century. Sometimes it has worked, sometimes it hasn't.
- Naturalness is not a mathematical criterion. It is a mathematically formulated beauty requirement. Its lack of success does not justify its use even as experience-based.

5
Ideal Theories

In which I search for the end of science but find that the imagination of theoretical physicists is endless. I fly to Austin, let Steven Weinberg talk at me, and realize how much we do just to avoid boredom.

Surprise Me—but Not Too Much

You might be surprised to hear how much Bach has in common with the Beatles.

In 1975, Richard Voss and John Clarke, two physicists from Berkeley, studied noise in electronic devices.[1] For the fun of it, they then applied their method to different types of music. Surprisingly, they found that different types of music—Western or Eastern, blues, jazz, or classical—all have a common pattern: while the loudness and pitches themselves differ between the music styles, the amount of variation decreases universally with the inverse of the frequency, or what's called a "1/*f* spectrum." This means long variations are rarer, but they have no preferred length—they can be of any duration.

The 1/*f* spectrum has—theoretically—no typical time scale to it, contrary to the expectation that verse meters or beats mark the type of music. The study therefore reveals that sound patterns in music have self-similarities or "correlations" that stretch over all time scales. White noise would show a constant spectrum and no correlation between fluctuations. A random walk moving a melody along adjacent pitches would have a strong correlation and a $1/f^2$ spectrum. Somewhere in between, Voss and Clarke showed, are Bach, the Beatles, and everything else you hear on the radio.[2]

Intuitively this means that good music lives on the edge between predictability and unpredictability. When we turn on the radio, we want to be surprised—but not too much. Not so surprisingly, then, popular music follows quite simple recipes, and you can sing along when the chorus repeats.

This observation about music, I think, carries over to other areas of human life. In the arts, in writing, and in science too we like to be surprised, but not too much. Though it's harder to quantify than sound patterns, scientific papers also need to strike a balance between old and new. Novelty is good, but not if it expects too much of the audience. The real pop stars, like the scientific pop stars, are those living just on the edge; they are the ones who make us slap our heads and mutter, "Why didn't I think of that?"

But ideas in science, unlike in the arts, aren't an end unto themselves; they are means to the end of describing the natural world. In science, new data can enforce change. But what if there are no new data? Then we reinvent hits from the past, in more or less obvious ways. And the new theories in physics, like new pop songs, remain variations on already familiar themes.

ꝏꝏ

IN THEORETICAL physics, the presently popular themes are simplicity, naturalness, and elegance. These terms are never defined, strictly speaking, and I won't attempt to define them either; I will just tell you how they are used.

Simplicity

Simplicity means doing with less. But as Einstein already observed, a theory should be "as simple as possible and not any simpler." The requirement of simplicity cannot by itself be used for theory development, for there are many theories simpler than the ones that describe our universe. To begin with, there isn't any good reason for there to be a universe at all, or for a universe to contain matter. Or, to use a less nihilistic example, quantizing gravity is considerably simpler in two dimensions—we just don't inhabit such a universe.

Simplicity therefore is merely of relative value. We can look for a theory simpler *than* some other theory, but not start constructing a theory based on simplicity alone.

It is almost tautologically true that of two theories that accomplish the same thing, scientists will eventually settle on the simpler one, because why would we want to make our life more difficult than necessary? Historically, there has sometimes been a delay in this settlement when simplicity was in conflict with other cherished ideals, such as the beauty of circular planetary motion. But laziness has always won, at least so far.

It's only *almost* tautological because simplicity is in a constant tug-of-war with accuracy. Additional parameters, and hence less simplicity, normally allow a better fit of data, and we can use statistical measures to calculate whether an improved data fit justifies a new parameter. One can debate the pros and cons of different measures, but for our purposes it suffices to say that the search for theory extensions that might conflict with simplicity is pursued in an area called phenomenology.[3]

A way to objectively quantify simplicity is by computational complexity, which is measured by the length of a computer program that executes a calculation. Computational complexity is in principle quantifiable for any theory that can be converted into computer code, which includes the type of theories we currently use in physics. We are not computers, however, and computational complexity is therefore not a measure we actually use. The human idea of simplicity is instead very much based on ease of applicability, which is closely tied to our ability to grasp an idea, hold it in mind, and push it around until a paper falls out.

To achieve simplicity for newly conjectured laws of nature, theorists presently try to minimize the sets of assumptions. This can be done by reducing the number of parameters, the number of fields, or, more generally, the number of axioms of a theory. The most common ways to do this are currently unification and the addition of symmetries.

That a fundamental theory shouldn't have unexplained parameters was also one of Einstein's dreams:

> *Nature is constituted so that it is possible to lay down such strongly determined laws that within these laws only rationally, completely determined constants occur (not constants, therefore, that could be changed without completely destroying the theory).*[4]

This dream still drives research today. But we do not know whether more fundamental theories necessarily have to be simpler. The assumption that a more fundamental theory should also be simpler—at least perceptually simpler—is a hope and not something that we actually have reason to expect.

Naturalness

In contrast to simplicity, which scrutinizes the number of assumptions, naturalness evaluates the type of assumptions. It is an attempt to get rid of the human element by requiring that a "natural" theory should not use cherry-picked assumptions.

Technical naturalness is distinct from general naturalness in that it applies only to quantum field theories. But both have a common basis: an assumption that is unlikely to have occurred by chance should be avoided.

The naturalness criterion, however, is useless without further assumptions, assumptions that require making an unexplained choice and thereby bring back cherry-picking. Problem is, there are infinitely many different ways for something to be due to chance, and so the reference to chance itself already requires a choice.

Consider this example. If you have a regular die, the probability for any one number to come up is the same: 1/6. But if you have an oddly shaped die, the probability could be different for each number. The oddly shaped die, we say, has a different "probability distribution," that is, a function that encodes the probability for each possible outcome to occur. This could be any function, so long as all probabilities add to 1.

When we say something is random without adding a qualifier, we usually mean it's a uniform probability distribution, that is, a distribution with equal probability for each outcome, like the regular die. But why should the probability distribution for the parameters of a theory be uniform? We have only one set of parameters that describe our observations. It's like someone told us the outcome of throwing a die once. But this doesn't tell us anything about the shape of the die. The uniform distribution, like the regular die, might seem pretty. But that's exactly the kind of human choice that naturalness attempts to get rid of.*

Worse, even if you cherry-pick a probability distribution, naturalness remains a meaningless criterion because it immediately renders unnatural all theories we can possibly conceive of. The reason is that naturalness requirements are at present selectively applied only to one type of assumption: dimensionless numbers. But in theory development we use many other assumptions that are selected "merely" to explain observations. We just don't normally talk about this.

An example is the stability of the vacuum. This is a common assumption that ensures that the universe around us doesn't spontaneously fall apart and tear us into pieces. Reasonable enough. But there are infinitely many "bad" theories in which this can happen. These theories aren't bad because they are mathematically wrong; they are bad simply because they wouldn't describe what we see. Vacuum

* For a more technical version of this argument, see Appendix B.

stability is chosen only for the purpose of describing nature, and yet nobody ever complains it is somehow cherry-picked and "unnatural." There are many other such assumptions that we select simply because they work, not because they are probable in some way. And if we're willing to accept all these other assumptions "just because," why not accept picking a parameter?

"Oh well," you might say, "we have to start somewhere. So let's start with explaining the parameters first and get to the more complicated assumptions later."

Look, I reply, the mere attempt of trying to justify why we use exactly these assumptions is logical mush: if you do not approve of selecting assumptions by means other than mathematics, then the only allowed requirement for a physical theory is mathematical consistency. Consequently, all logically consistent sets of axioms are equally good, and there are infinitely many of these. But that's entirely useless to describe nature—we don't want to just list consistent theories, we want to explain our observations. And for that, we necessarily need to compare prediction with observation to select useful assumptions for our theories. Which is what we did before we got carried away by ideals like naturalness.

Neither is the idea that numbers close to 1 are somehow preferable rooted in mathematics. If you dig a bit in obscure areas of mathematics, you will find numbers of all sizes and shapes, according to taste. One stunning example is the number of elements of what is aptly called the "monster group." It comes to 808,017,424,794,512, 875,886,459,904,961,710,757,005,754,368,000,000,000.

That's approximately 10^{54}, in case you don't feel like counting digits. Luckily, no number of that size currently needs explaining in physics; otherwise I'm sure someone would have tried to use the monster group for it.

So no, we can't blame math for our love of numbers that feel good.

Don't get me wrong—I agree it's generally preferable to have a better explanation for any assumption that we make. I merely object to the idea that some numbers are in particular need of explanation while other problems fall by the wayside.

I hasten to add it's not as though naturalness is always useless. It can be applied if we know the probability distribution—for example, the distribution of stars in the universe, or the distribution of fluctuations in a medium. We can then tell what is and isn't a "natural" distance to the next star or a "likely" event. And if we have a theory that, like the standard model, upon inspection turns out to have many natural parameters, it's reasonable to extrapolate this regularity and base predictions on it. But if the prediction fails, we should note that and move on.

In practice the dominance of naturalness means you won't be able to convince anyone to do an experiment without an argument for why new physics is "naturally" supposed to appear in the experiment's reach. But since naturalness is fundamentally aesthetic, we can always come up with new arguments and revise the numbers. This has resulted in decades of predictions for new effects that were always just about measurable with an upcoming experiment. And if that experiment didn't find anything, the predictions were revised to fall within the scope of the next upcoming experiment.

Elegance

Finally, there is elegance, the most elusive of the criteria. It is often described as a combination of simplicity with surprise that, taken together, reveals relevant new knowledge. We find elegance in the aha effect, the moment of insight when everything falls into place. It's what philosopher Richard Dawid referred to as "unexpected explanatory closure"—the unanticipated connection between the formerly unrelated. But it's also the simple giving rise to the complex, opening a new vista on untouched ground, a richness in structure that—surprisingly—grows out of economy.

Elegance is an unabashedly subjective criterion, and while it is influential, no one has tried to formalize and use it for theory development. So far. Richard Dawid's is the first attempt to define a sense of elegance through unexpected explanatory closure in his proposed method of theory assessment. But to the extent that it's an explanatory closure, it's a demand for consistency, which is a quality requirement anyway. And to the extent that the closure should be "unexpected," it's a statement about the ability of the human brain to anticipate mathematical results before they've been derived. Hence it remains a subjective criterion.

Beauty, then, is a combination of all of the above: simplicity, naturalness, and a dose of the unexpected. And we play by these rules. After all, we don't want to surprise anybody too much.

∞∞

THE MORE I try to understand my colleagues' reliance on beauty, the less sense it makes to me. Mathematical rigidity I had to discard because it rests on the selection of a priori truths, a choice that is itself not rigid, turning the idea into an absurdity. Neither could I find a mathematical basis for simplicity, naturalness, or elegance, each of which in the end brought back subjective, human values. In using these criteria, I fear, we overstep the limits of science.

Someone needs to talk me out of my growing suspicion that theoretical physicists are collectively delusional, unable or unwilling to recognize their unscientific procedures. I need to talk to someone who has had the experience that these criteria work, experience that I lack. And I know just the right person for that.

Breeding Horses

It's January and it's Austin, Texas. I'm so scared of being late that I arrive an hour early for my appointment with Steven Weinberg. The hour calls for a very big coffee. When I'm halfway through my cup, a young man comes by and asks if he can sit with me. Sure, I say. He places a big book on the table and starts going through the text, taking notes. I glance at the equations. It's classical mechanics, the very beginning of first-semester theoretical physics.

A crowd of students walks by, chatting. I ask the studious youngster if he knows which lecture they might have come from. "No, sorry," he says, and adds he's only been here for two weeks. Has he decided, I ask, which area of physics he wants to specialize in? He tells me he read Brian Greene's books and is very interested in string theory. I tell him that string theory isn't the only game in town, that physics isn't math, and that logic alone will not suffice to find the right theory. But I don't think my words count much against Greene's.

I give the young man my email address, then stand up and walk down a windowless corridor, past withered conference posters and seminar announcements, until I find a door plate that reads "Prof. Steven Weinberg." I peek in but the professor hasn't arrived yet. His secretary ignores me, and so I wait, watching my feet, until I hear steps in the corridor.

"I'm supposed to speak to a writer now," Weinberg says, and looks around, but there's only me. "Is that you?"

Always keen on new opportunities to feel entirely inadequate, I say yes, thinking I shouldn't be here, I should be at my desk, reading a paper, drafting a proposal, or at least writing a referee report. I shouldn't psychoanalyze a community that neither needs nor wants therapy. And I shouldn't pretend to be something I'm not.

Weinberg raises an eyebrow and points to his office.

His office, it turns out, is half the size of mine, an observation that vaporizes what little ambition I ever had to win a Nobel Prize. I don't have, of course, all those honorary titles on the wall. Neither do I have my own books to line up on my desk. Weinberg has now made it up to a dozen.

His *Gravitation and Cosmology* was the first textbook I ever purchased to keep. It was so shockingly expensive that for the better part of a year I dragged the book around with me wherever I went, afraid I might misplace it. I went to the gym with the book. I ate over the book. I slept with the book. I even eventually opened it.

The book has a plain, dark blue cover with golden imprint; it basically begs for a layer of dust. Imagine how excited I was when I noticed that the author was still alive and not, as I had assumed, a long-deceased contemporary of Einstein and Heisenberg, the men who had so far made up most of my literature. Indeed, not only was the author still alive, but in the following years he would go on to publish three volumes on quantum field theory. I slept with them too.

Now in his mid-eighties, Weinberg is still doing research and still writing books, with a new one just coming out. If there's anyone on this planet able to tell me why I should rely on beauty and naturalness in my research, it's him. I grab my notepad, take a seat, and hope I look writerly enough.

So, I think as I turn on the recorder, I finally get to ask about the damned horse breeder.

"You have this analogy with the horse breeder. That seems to suggest that paying attention to beauty in theory construction is based on experience?"

"Yes, I think it is," Weinberg says. "If you go back to the Greeks of the Hellenic period up to the time of Aristotle..."

Weinberg doesn't talk with you, they told me, he talks *at* you. Now I know what they mean. And let me tell you, he talks like a book, almost print-ready.

"If you go back to the Greeks of the Hellenic period, up to the time of Aristotle, they seemed to feel that they had an innate sense of rightness which had a moral quality to it. For example, Parmenides had an incredibly simple theory of nature that nothing ever changes. It's contradicted by experience, but he never went to the trouble of reconciling appearances with his theory of unchangeableness. He asserted this theory merely on grounds of simplicity and elegance and, I think, a kind of snobbery that change is always less noble than permanence.

"We've learned to do better than that. We've learned that whatever theories are suggested by our aesthetic ideas, they somehow have to be confronted by actual experience. And as time has passed we have learned not only that the theories suggested by our sense of beauty have to be confirmed by experience, but gradually our sense of beauty changes in a way that is produced by our experience.

"For example, that wonderful conception of a holistic view of nature, that the same values that affect human life—love and hate and strife and justice and so on—somehow can be applied to the inanimate world, this holistic picture of nature, which is exemplified by astrology, that what happens in the heavens has a direct correlation with what happens to human beings—that used to be thought very beautiful, because it was a unified theory of everything. But we've learned to give it up. We don't look for human values in the laws of nature any longer. We don't talk about noble elementary particles, or scattering in a nuclear reaction occurring in order to reach a 'just' result in the way Anaximander did.

"Our sense of beauty has changed. And as I described it in my book, the beauty that we seek now, not in art, not in home decoration—or in horse breeding—but the beauty we seek in physical theories is a beauty of rigidity. We would like theories that to the greatest extent possible could not be varied without leading to impossibilities, like mathematical inconsistencies.

"For example, in the modern standard model of elementary particles, we have six kinds of quarks and six kinds of leptons.* So they're equal in number. You could say that's very pretty that there is a one-to-one correspondence. But what's really pretty about it is that if there *wasn't* a one-to-one correspondence, you wouldn't have mathematical consistency. The equations would have a problem which is known in the trade as an 'anomaly,' and you wouldn't have consistency. Within the general formalism of the standard model you have to have equal numbers of quarks and leptons.

"We find that beautiful because, for one thing, it satisfies our desire for explanation. Knowing that there are six types of quarks, we can understand why there are also six types of leptons. But also it satisfies our sense that what we're dealing with is something that is almost inescapable. The theory we have is one that explains itself. Not entirely—we don't know why the number is six rather than four or twelve or any other even number—but to some extent it explains itself in terms of mathematical consistency. And that's wonderful because it moves us further along in explaining the world, to use the title of my latest book."[5]

* Leptons are the standard model fermions that aren't quarks.

In brief, he says, we're so much smarter now than they were then because math prevents us from making vague or mutually contradictory assumptions. In his story, theoretical physicists are scientific demigods, getting ever closer to their dream of a final theory. Oh, how I want to believe it! But I can't. And since my lapse of faith is what brought me here, I object to the idea of a theory explaining itself.

"But mathematical consistency seems a fairly weak requirement," I say. "There are many things that are mathematically consistent that have nothing to do with the world."

"Yes, that's right," Weinberg says. "And my guess is that in the end mathematical consistency will not be a strong enough requirement to identify a single possible theory. I think the best we can hope for is a theory that is unique in the sense that it is the only mathematically consistent theory that is rich, that has lots and lots of phenomena, and in particular that makes it possible for life to arise.

"You see," he continues, "I am sure you are right that mathematical consistency is not enough, because we can invent theories that we think are mathematically consistent that certainly don't describe the real world. Like a theory with just one particle, not interacting with anything, sitting in an empty space, nothing ever happening—that's a mathematically consistent theory. But it's not very interesting, or rich. And maybe the real world is governed by the only mathematically consistent theory that allows a rich world, lots of phenomena, lots of history. But we're very far from reaching this conclusion."

Imagine we succeeded in this. Imagine theoretical physicists proved there is only one, ultimate law of nature that could have created us. Finally, everything would make sense: stars and planets, light and dark, life and death. We would know the reason for each and every happenstance, know that it could not have been any different, could not have been better, could not have been worse. We'd be on par with nature, able to look at the universe and say, "I understand."

It's the old dream of finding meaning in the seemingly meaningless. But it's not just about making sense. Armed with this breakthrough, theoretical physicists would become the arbiters of truth. Certain now how to uphold natural law, they would take on the rest of science, releasing the insights still locked in mystery. They would change the world. They would be heroes. And they'd finally be able to calculate the mass of the Higgs boson.

I can see the appeal. I cannot, however, see that deriving a unique law of nature is more than a dream. To get there, I don't think it's any good to use naturalness for guidance. And what is the alternative? Weinberg expressed his hope that there is only one "mathematically consistent theory that allows a rich world." But finding a consistent theory that is not in conflict with observation is easy: just use a theory that doesn't make predictions. This can't be what he meant, I think, and so I say, "That would require, though, that the theory is predictive enough so that it gives rise to the parameters necessary for complex atomic and nuclear physics."

"I don't know," Weinberg says with a shrug. "It might be that the correct theory allows a great number of different big bangs to evolve out of the early stages of the universe. In these different big bangs, all of which are going on at the same time, the constants of nature are very different, and we can never predict what they are, because whatever they are, they're only that in our big bang. It would be like trying to predict, from first principles, the distance of Earth from the Sun. Obviously there are billions of planets and they're all at different distances from their stars and it's something we can't ever hope to predict. And it may be also be that there is an unlimited number of big bangs. And the values of constants of nature are just whatever they happen to be in the big bang we're inhabiting.

"These are wild speculations," Weinberg says. "We don't know that any of that is true. But it's certainly a logical possibility. And there are physical theories in which it would be true."

Endless Possibilities

You're one human among some 7 billion on this planet; your Sun is one star among a hundred billion in the Milky Way; the Milky Way is one galaxy among some 100 billion in the universe. Maybe there are other universes making up what we call the "multiverse." Doesn't sound like such a big deal? It's presently the most controversial idea in physics.

Cosmologist Paul Steinhardt calls the multiverse idea "baroque, unnatural, untestable and, in the end, dangerous to science and society."[6] According to Paul Davies, it is "simply naive deism dressed up in scientific language."[7] George Ellis warns that "proponents of the multiverse . . . are implicitly redefining what is meant by 'science.' "[8] David Gross finds that "it smells of angels."[9] For Neil Turok, it's "the ultimate catastrophe."[10] And science writer John Horgan complains that "multiverse theories aren't theories—they're science fictions, theologies, works of the imagination unconstrained by evidence."[11]

On the other side of the controversy we have Leonard Susskind, who finds it "exciting to think that the universe may be much bigger, richer and [more] full of variety than we ever expected."[12] Bernard Carr reasons that "the notion of a multiverse entails a new perspective of the nature of science and it is not surprising that this causes intellectual discomfort."[13] Max Tegmark argues that multiverse opponents have an "emotional bias against removing ourselves from center stage."[14] And Tom Siegfried thinks the critics have "the same attitude that led some 19th century scientists and philosophers to deny the existence of atoms."[15] Ouch.

So what's the big deal? It's that Einstein taught us nothing can travel through space faster than light. This means that at any given moment the speed of light sets a limit to how far we can see, a limit known as the "cosmological horizon." Any messenger other than light would be slower, or—in the case of gravity itself—equally fast as light. Therefore, if something is so far away that light from it couldn't yet have reached us, we would not be able to tell it's there at all.

But while nothing can travel through space faster than light, space itself knows no such limit. It can and, in the multiverse, does expand faster than light, and so there are regions from which light can never reach us. In the multiverse, all the other universes are in such regions and therefore causally disconnected from us. Out of reach, eternally. Hence, say the multiverse opponents, you can't ever measure it, so it's not in the realm of science.

In return, proponents of the multiverse point out that just because a theory has elements that are unobservable, that doesn't mean the theory cannot make predictions. We have known since the dawn of quantum mechanics that it's wrong to require all mathematical structures of a theory to directly correspond to observables. For example, wave functions are not themselves measurable; what is measurable is merely the probability distribution that is derived from the wave function. Now, not everybody is happy with this state of affairs. But we all agree that quantum mechanics is highly successful regardless.

How willing physicists are to accept non-observable ingredients of a theory as necessary depends on their trust in the theory and on their hope that it might give rise to deeper insights. But there isn't a priori anything unscientific about a theory that contains unobservable elements.

Extracting predictions from a multiverse might be possible, despite most of it being unobservable, by studying the probability that one of the universes in the multiverse has laws of nature like ours. We would then not be able to actually derive the fundamental laws of nature in our universe, but we might still be able to conclude which laws we would most likely observe. And that, claim the idea's proponents, is the best we can do. It's a paradigm change, a shift in perspective about what it means for a statement to be scientific to begin with. If you're not buying into it, if you're not accepting the new science, you're blocking progress and are hopelessly behind, a fossil ready to be buried in sediment.

You can't calculate any probabilities in the multiverse, say the opponents of the idea, because there are infinitely many instances of all possibilities, and you can't meaningfully compare infinities to other infinities. It is possible, but for this you need a mathematical scheme—a probability distribution, or "measure"—that tells you how to tame the infinities. And where does that probability distribution come from? For that you need another theory, and at that point you might as well try to find a theory that doesn't produce all these unobservable universes to begin with.

It's not optional, the multiverse proponents reply. If we live in the best of all possible worlds, then what about the rest of all possible worlds? They can't just be ignored. We haven't cooked this up, they say; our theories force us to eat it. It's not us; it's the math that made us do it. And math doesn't lie. We are merely being objective, good scientists, they say. If you're opposing these insights, you are in denial and just refuse to accept inconvenient logical consequences.

And so it's been going for two decades.

OOOO

WE HAVE no reason to believe that the universe ends beyond the cosmological horizon and that starting tomorrow we'll discover that there are no more galaxies beyond what we can see today. But just how far the distribution of galaxies similar to the ones around us continues, nobody knows—nobody can know. We don't even know whether space continues indefinitely or whether it eventually closes back onto itself, giving rise to a finite, closed universe with a radius much larger than what we presently see.

This continuation of the universe as we know it is uncontroversial and not what is normally meant by "multiverse." A multiverse proper instead contains regions that look nothing like what we observe around us. And there are various types of such multiverses that theoretical physicists today believe are logical consequences of their theories:

1. Eternal Inflation

Our understanding of the early universe is limited because we don't know much about matter at temperatures and densities higher than what we have so far been able to probe. We can, however, extrapolate our theories—the standard model and the concordance model—by assuming they continue to work just the same way. And if we extrapolate the behavior of matter to earlier and earlier times, we must extrapolate the behavior of space-time along with it.

Currently the most widely accepted extrapolation into the past has it that the universe once went through a rapid phase of expansion, known as "inflation." Inflation is caused by a new field, the "inflaton," whose effect is to speed up the universe's expansion. The inflaton blows up the universe like dark energy, but does so much faster. When inflation ends, the inflaton's energy is converted into particles of the standard model and dark matter. From there on, the history of the universe continues as we discussed in Chapter 4.

Inflation has some evidence speaking for it, though not overwhelmingly so. Nevertheless, physicists have further extrapolated this extrapolation to what is known as "eternal inflation." In eternal inflation, our home universe is just a small patch in a much larger—indeed, infinitely large—space that inflates, and will continue to inflate forever. But because the inflaton has quantum fluctuations, bubbles where inflation ends can appear, and if such a bubble becomes large enough, galaxies can form in it. Our universe is contained in such a bubble. Outside our bubble, space still inflates, and randomly occurring quantum fluctuations spawn other bubble universes—eternally. These bubbles form the multiverse. If one believes that this theory is correct, say the multiverse proponents, then the other universes must be just as real as ours.

The idea of eternal inflation is almost as old as that of inflation itself; it showed up as an unintended side effect of the first inflationary models in 1983.[16] But eternal inflation got little attention until the mid-1990s, when string theorists discovered its use. Today, most of the inflationary models under study result in a multiverse, though not all of them do.[17]

In the multiverse of eternal inflation, with an infinite number of recurrences launched by random quantum fluctuations, anything that can happen will eventually happen. Eternal inflation therefore implies that there are universes in which the history of mankind plays out in any way that is compatible with the laws of nature. In some of them this will make sense to you.

2. The String Theory Landscape

String theorists hoped to uncover a theory of everything that would contain both the standard model and general relativity. But starting in the late 1980s, it became increasingly clear that the theory cannot predict which particles, fields, and parameters we have in the standard model. Instead, string theory gives rise to a whole landscape of possibilities. In this landscape every point corresponds to a different version of the theory with different particles and different parameters and different laws of nature.

If one believes that string theory is the final theory, then this lack of predictability is a big problem: it means the theory cannot explain why we observe this particular universe. Hence, to make the final theory claim consistent with the lack of predictability, string theorists had to accept that any possible universe in the landscape has the same right to existence as ours. Consequently, we live in a multiverse. The string theory landscape conveniently merged with eternal inflation.

3. Many Worlds

The many-worlds interpretation is a variant of quantum mechanics (more about this in Chapter 6). In this interpretation, rather than just one of the outcomes for which quantum mechanics predicts probabilities being realized, all of them are, each in its own universe. We arrive at this picture of reality by removing the assumption that the measurement process in quantum mechanics singles out one particular outcome. Again we see that it's a reliance on mathematics together with a desire for simplicity that leads to multiple universes.

The many worlds of quantum mechanics are distinct from the different universes in the landscape, since the existence of many worlds doesn't necessarily imply a change of the particle types from one universe to the next. It is possible, however, to combine these different multiverses into an even larger one.[18]

4. The Mathematical Universe

The mathematical universe takes the final-theory claim to its extreme. Since any theory that describes our universe requires the selection of some mathematics among all possible mathematics, an ultimate final theory cannot justify any particular selection because that would require yet another theory to explain it. Therefore, the logical conclusion must be that the only final theory is one in which all mathematics exists, making up a multiverse in which the square root of –1 is just as real as you and me.

This idea of the mathematical universe, in which all these mathematically possible structures exist, was put forward by Max Tegmark in 2007.[19] It subsumes all other multiverses. While it enjoys a certain popularity among philosophers, most physicists have ignored it.

This list may raise the impression that the multiverse is news, but the only new thing here is physicists' insistence on believing in its reality. Since every theory requires observational input to fix parameters or pick axioms, every theory leads to a multiverse when it lacks input. For then you could say, "Well, all these possible choices must exist, and therefore they make up a multiverse." If the goal is a theory that can explain everything from nothing, then the predictions of that theory must eventually become ambiguous—and so the multiverse is unavoidable.

Newton, for example, could have refused to just measure the gravitational constant and instead argued that there must be a universe for each possible value. Einstein could have argued that all solutions to the equations of general relativity must exist somewhere in the multiverse. You can create a multiverse for every theory—all you have to do is to drop a sufficient amount of assumptions or ties to observation.

The origin of the multiverse fad, therefore, is that some physicists are no longer content with a theory that describes observation. In trying to outdo themselves, they get rid of too many assumptions, and then they conclude we live in a multiverse because they can't explain anything anymore.

But maybe I am a fossil, ready to be buried in sediment.

Science has the goal of describing our observations of nature, but a theory's predictions may have a long incubation period before they hatch. And so, I contend, taking multiverses seriously, rather than treating them as the mathematical artifacts that I think they are, might eventually lead to new insights. The question then boils down to how plausible it is that new insights will emerge from this approach, and that returns me to the problem that sent me on this trip: how do we assess the promise of a theory that has no observational evidence speaking for it?

Not all variants of the multiverse are entirely untestable. In eternal inflation, for example, our universe could have collided with another universe in the past, and such a "bubble collision" could have left an observable signature in the cosmic microwave background. It's been looked for and hasn't been found. This doesn't rule out the multiverse, but it rules out that there has been such a collision.

Other physicists have argued that some multiverse variants might give rise to a distribution of small black holes in our universe, which has consequences that might become observable soon.[20] But a multiverse prediction that doesn't come true merely means that we need a probability distribution to make the non-observed phenomenon unlikely, so let's look for a distribution that works. This approach to cosmology is as promising as trying to understand *War and Peace* by discarding every other book. But, I commend, at least they're trying.

The existing predictions demonstrate that the multiverse is in principle amenable to experimental test, but these tests are useful only for very specific scenarios. The vast majority of multiverse ideas are presently untestable, and will remain so eternally.

Thus prohibited from cranking the wheels of science, some theoretical physicists have proceeded by their own method of theory assessment: betting. Cosmologist Martin Rees bet his dog that the multiverse is right, Andrei Linde bet his life, and Steven Weinberg had "just enough confidence about the multiverse to bet the lives of *both* Andrei Linde and Martin Rees's dog."[21]

Cosmic Poker

The multiverse has gained in popularity while naturalness has come under stress, and physicists now pitch one as the other's alternative. If we can't find a natural explanation for a number, so the argument goes, then there isn't any. Just choosing a parameter is too ugly. Therefore, if the parameter is not natural, then it can take on any value, and for every possible value there's a universe. This leads to the bizarre conclusion that if we don't see supersymmetric particles at the LHC, then we live in a multiverse.

I can't believe what this once-venerable profession has become. Theoretical physicists used to explain what was observed. Now they try to explain why they can't explain what was not observed. And they're not even good at that. In the multiverse, you can't explain the values of parameters; at best you can estimate their likelihood. But there are many ways to not explain something.

"So," I say to Weinberg, "in the case that we live in a multiverse, the requirement that the theory is interesting enough, or gives rise to interesting enough physics, isn't this an empty requirement? You could do the same thing with any theory that doesn't predict the parameters."

"Well, you have to predict some things," Weinberg says. "Even if you don't predict the parameters, you may predict correlation between them. Or you may predict the parameters in terms of some theory, like the standard model but a more powerful theory which actually tells you the mass of the electron and so on, and then if you ask, 'Why is that theory correct?' you say, 'Well, that's the farthest we can go, we can't go beyond that.'"

He continues: "I wouldn't be in a hurry to set clear requirements for what a good theory has to be. But I can certainly tell you what a *better* theory has to be. A theory better than the standard model would be one that makes it inevitable that you have six rather than eight or four quarks and leptons. There are many things in the standard model that seem arbitrary, and a better theory would be one that makes these things less arbitrary, or not arbitrary at all.

"But we don't know how far we can go in that direction," he continues. "I don't know how much elementary particle physics can improve over what we have now. I just don't know. I think it's important to try and continue to do experiments, to continue to build large facilities. . . . But where it will wind up I don't know. I hope it doesn't just stop where it is now. Because I don't find this entirely satisfying."

"Has the recent data that has come from the LHC changed your mind about what this better theory looks like?" I ask.

"No, unfortunately it has not," Weinberg says. "That's been a huge disappointment. There's some hint that there might be some new physics at energies six times higher than the mass of the Higgs particle.* And that would be wonderful. But we were hoping that the LHC would reveal something really new. Not just continue to confirm the standard model, but find signs of dark matter or supersymmetry or something that would lead to the next big step in fundamental physics. And that hasn't happened."

"My understanding," I say, "was that a lot of this hope that the LHC should find something new was based on naturalness arguments, that for the Higgs mass to be natural there should also be new physics in this range."

"I don't take seriously any negative conclusion that the fact that the LHC hasn't seen anything beyond the standard model shows that there isn't anything that will solve the naturalness problems. All you can do is rule out specific theories, and we don't have any specific theory that is attractive enough so you can say you have really accomplished something when you've ruled it out. Supersymmetry hasn't been ruled out because it's too vague about what it predicts."

"If there is a fundamental theory, do you think it has to be natural?" I inquire.

"Yes," Weinberg says, "because if it wasn't, we wouldn't be interested in it." He goes on to explain: "By natural I don't mean some technical definition of naturalness by some particle theorists that are trying to constrain existing theories. By naturalness I mean that there is no completely unexpected equality or enormous ratios."

He pauses for a moment, then continues, "But maybe I answered too quickly when I said that a successful theory has to be natural,

* That's the diphoton anomaly.

because some things it might leave unexplained: they might be purely environmental in a multiverse where you have many different universes. It's just like astronomers used to think that a successful theory of the solar system would be one that made it natural to have Mercury, Mars, and Venus to be where they are. And Kepler tried to construct such a theory based on a geometric picture that involved Platonic solids.

"But we now know that we shouldn't look for such a theory because there is nothing natural about the distances of the planets from the Sun: they are what they are because of historical accidents." However, he adds, there are cases "like the fact that the period of rotation of Mercury is two-thirds of its orbital period. That number is explained in terms of the tidal forces of the Sun acting on Mercury.

"So some things can be explained. But in general what you see in the solar system is just a historical accident. And this natural explanation which astronomers like Kepler hoped for, or Claudius Ptolemy before him, you have to give up this search. It's just what it is." He pauses, then adds, "I hope that won't be so for the mass of the electron."

"So naturalness, leaving aside the technical definition, means that there are no unexplained parameters?" I ask.

"There are things that cry out for an explanation," Weinberg says. "Like a 2-to-1 ratio, or something being 15 orders of magnitude smaller than something else. When you see that, you feel like you have to explain that. And naturalness just means that the theory has the explanations to those things—they aren't just put in to make it agree with experiment."

"That's also an experience value?"

"Oh, yes," Weinberg says. "There are some things that you expect to have a natural explanation, and other things you don't."

Then he gives an example: "If you are in a poker game and three times in a row you get a royal flush, you'd say, 'Well, that's something that needs to be explained in terms of what the dealer is trying to accomplish.' On the other hand, if you get three hands which most of us would call random—a king of spades, a two of diamonds, and so on—and each of the three is different from the others, there's nothing special; none of them are winning hands; you'd say there's nothing to be explained. We don't need a natural explanation for that; these are just as likely as any other hand. Well, a royal flush is just as likely or unlikely as any other hand. But there is still something about a royal flush that cries out for an explanation if you get three in a row."

In Weinberg's game, the cards are the laws of nature or, more specifically, the parameters in the laws we presently use. But it's a game we never played, never could have played. We were handed a set of cards, not knowing why or by whom. We have no idea how likely we were to get the royal flush of natural law, or if there is anything special to it. We don't know the rules of the game, and we don't know the odds.

"It depends on the probability distribution," I say, trying to explain my dilemma: that any such statement about the likelihood of the laws of nature needs yet another law, one for the likelihood of the laws. And simplicity would then prefer the theory with just a fixed parameter, rather than a probability distribution of that parameter over a multiverse.

"Well," Weinberg repeats, "the probability of getting a two of clubs, a five of diamonds, a seven of hearts, an eight of hearts, and a jack of hearts, the probability of getting that particular hand, is exactly the same as the probability of getting ace, king, queen, jack, ten all of spades. The two have the same probability."

Trying to find a poker metaphor for the probability distribution, I say, "Provided that the dealer was fair." But the anthropomorphic example makes me uneasy. I can't shake the impression that we're really trying to guess the rules God plays by, in order to make sure the laws of nature were chosen fairly, hoping that maybe God made a mistake and we deserve a universe with low-mass gluinos.

"Yes," Weinberg replies to my comment, going on with the poker metaphor. "But because of human values that are associated with the different hands of pokers—that one of them will win and another one won't, because those are the rules of poker—you begin to pay attention when one of the people you're playing with gets a royal flush, in a way that you don't when he gets some perfectly ordinary hand, which actually is just as improbable as a royal flush. It's a human attribute to the royal flush that we say, 'Well, that's a winning hand.' And so it attracts your attention."

Right, it's a human attribute that coincidences attract our attention, like gauge-coupling unification or sliced bread that pops out of the toaster with an image of the Virgin Mary. But I can't see why this human attribute is of use in developing better theories.

"I am using this example to agree with you," Weinberg says, to my befuddlement. "If you didn't know anything about the rules of poker, you might not know that there was something special about a royal flush compared to any other hand. It's because we know the rules of poker that they seem special. It's a matter of experience value."

But we don't have any experience in cosmic poker! I think, distressed because I still can't see what any of this has to do with science. We have no way to tell whether the laws of nature we observe are likely—no probability distribution, no probability. In order to be able to determine that the laws are unlikely, we'd need another theory, and where did *that* theory come from?[22] If Weinberg, whom I count as the greatest physicist alive, can't tell me, then who can? So I ask again: "What then do we know about the probability distribution of these parameters?"

"Ah, well, you need a theory to calculate it."

Exactly.

ooOo

TO CALCULATE probabilities in the multiverse we must take into account that life exists in our universe. It sounds obvious, but not every possible law of nature creates sufficiently complex structures, and therefore the correct law must fulfill certain requirements—for example, give rise to stable atoms, or something similar to atoms. This requirement is known as the "anthropic principle."

The anthropic principle doesn't normally result in precise conclusions, but within the context of a specific theory it lets us estimate which values the theory's parameters can possibly have and still be compatible with the observation that life exists. It's like when you see someone walking down the street with a Starbucks cup and you conclude the conditions in this part of town must allow for Starbucks cups to arise. You might conclude the next Starbucks is within a one-mile radius, or maybe within a five-mile radius, and most likely closer than a hundred miles. Not very precise, and maybe not terribly interesting, but still, it tells you something about your environment.

While the anthropic principle might strike you as somewhat silly and trivially true, it can be useful to rule out some values of certain parameters. When I see you drive to work by car every morning, for example, I can conclude you must be old enough to have a driver's license. You might just be stubbornly disobeying laws, but the universe can't do that.

I have to warn you, however, that the reference to "life" in connection with the anthropic principle or fine-tuning is a common but superfluous verbal decoration. Physicists don't have a lot of business with the science of self-aware conscious beings. They talk about the formation of galaxies, the ignition of stars, or the stability of atoms, which are preconditions for the development of biochemistry. But don't expect physicists to even discuss large molecules. Talking about "life" is arguably catchier, but that's really all there is to it.

The first successful use of the anthropic principle was Fred Hoyle's 1954 prediction of properties of the carbon nucleus that are necessary to enable the formation of carbon in stellar interiors, properties that were later discovered as predicted.[23] Hoyle is said to have exploited the fact that carbon is central to life on Earth, hence stars must be able to produce it. Some historians have questioned whether this was indeed Hoyle's reasoning, but the mere fact that it could have been shows that anthropic argumentation can lead to useful conclusions.[24]

The anthropic principle, therefore, is a constraint on our theories that enforces agreement with observation. It is correct regardless of whether there is a multiverse or not, and regardless of the underlying explanation for the values of parameters in our theories—if there is an explanation.

The reason the anthropic principle is often brought up by multiverse proponents is that they claim it is the *only* explanation and there is no other rationale that selects the parameters we observe. One then needs to show, though, that the values we observe are indeed the only ones (or at least very probable) if one requires that life is possible. And this is highly controversial.

The typical claim that the anthropic principle explains the value of parameters in the multiverse goes like this: if parameter x was just a little larger or smaller, we wouldn't exist. There are a handful of examples for which this is the case, like the strength of the strong nuclear force: make it weaker and large atoms won't hold together, make it stronger and stars will burn out too quickly.[25] The problem with such arguments is that small variations in one out of two dozen parameters ignore the bulk of possible combinations. You'd really have to consider independent modifications of all parameters to be able to conclude there is only one combination supportive of life. But this is not presently a feasible calculation.

Though right now we cannot scan the whole parameter space to find out which combinations might support life, we can try at least a few. This has been done and is the reason we now think there is more than one combination of parameters that will create a universe hospitable to life.

For example in their 2006 paper "A Universe Without Weak Interactions," Roni Harnik, Graham Kribs, and Gilad Perez put forward a universe that seems capable of creating preconditions for life and yet has fundamental particles entirely different from our own.[26] In 2013, Abraham Loeb, of Harvard, argued that a primitive form of life might have been possible in the early universe.[27] And recently, Fred Adams and Evan Grohs showed that if we vary several parameters in our theories at once, there are ways for stars to produce carbon other than the mechanism Hoyle predicted—it's just that those other options conflict with observations that Hoyle already knew of.[28]

These three examples show that a chemistry complex enough to support life can arise under circumstances that are not anything like the ones we experience, and our universe isn't all that special.[29]

However, the anthropic principle might still work for some parameter if that parameter's effect is almost independent of what the other parameters do. That is, even if we cannot use the anthropic principle to explain the values of all parameters because we know there are other combinations allowing for the preconditions of life, certain parameters might need to have the same value in all cases because their effect is universal. The cosmological constant is often claimed to be of this type.[30] If it's too large, then the universe either will rip apart all the structures in it or will recollapse before stars have the chance to form (depending on the constant's sign).

Still, if we want to derive a probability rather than a constraint, we need a probability distribution for the possible theories, and that distribution can't come from the theories themselves—it requires additional mathematical structure, a metatheory that tells you how probable each theory is. It's the same problem that occurs with naturalness: the attempt to get rid of human choice just moves the choice elsewhere. And in both cases, one has to add unnecessary assumptions—in this case, the probability distribution—that could be avoided by just using a simple fixed parameter.

∞∞

THE MOST famous multiverse-based prediction is that of the cosmological constant, and it came from none other than Steven Weinberg.[31] The prediction dates back to 1997, when string theorists had just recognized they could not avoid getting a multiverse in their theory. Weinberg's paper significantly contributed to the acceptance of the multiverse as a scientific idea.

"I, together with two people here in the astronomy department, Paul Shapiro and Hugo Martel, we wrote a paper some years ago," Weinberg says. "We were concentrating on a particular constant, the cosmological constant, which has been measured in recent years through its effect on the expansion of the universe.*

* As recently as 1998.

"We assumed the probability distribution was completely flat, that all values of the constant are equally likely. Then we said, 'What we see is biased because it has to have a value that allows for the evolution of life. So what is the biased probability distribution?' And we calculated the curve for the probability and asked, 'Where is the maximum? What is the most likely value?'" The most likely value turned out to be quite close to the value of the cosmological constant, which was measured a year later.[32]

"So," Weinberg explains, "you could say if you had a fundamental theory that predicted a vast number of individual big bangs with varying values of the dark energy and an intrinsic probability distribution for the cosmological constant that is flat—that doesn't distinguish one value from another—then what living beings would expect to see is exactly what they see."

Or maybe, I think, you could say the multiverse is just a mathematical tool. Like those internal spaces. Whether it's really real is a question we can leave to philosophers.

I push this idea around for a while, but I still can't see how guessing a probability distribution for a parameter is any better than guessing the parameter itself. Except that you can publish the first guess because it involves a calculation, while the latter is just obviously nonscientific.

"I should say, we've been talking half an hour and my voice is giving in," Steven Weinberg says, coughing. "I have a feeling that I've covered just about everything I wanted to. Can we cut this short?"

The polite thing to do at this point, I think, is to say thank you and leave.

Emancipating Dissonance

I can't say I'm a fan of twelve-tone music. But I also admit I haven't spent much time listening to it. Someone who has done that is music critic Anthony Tommasini. In a 2007 video for the *New York Times* he speaks about the "emancipating dissonance" in the compositions of Arnold Schoenberg, the inventor of twelve-tone music.[33] Schoenberg's innovation dates back to the 1920s and enjoyed a brief period of popularity among professional musicians in the 1970s, but it never caught on more widely.

"Schoenberg would be very upset if you thought of his music as dissonant in a harsh, pejorative, negative way," Tommasini relates. "He thought he was allowing full life and richness and complexity.... For example, here is a piano piece from Opus 19 which is very dissonant, but it's delicate and gorgeous." He plays a few bars, then adds another example. "You can harmonize [this theme] in C-major," Tommasini demonstrates, clearly dissatisfied, "which is *so* boring in comparison with what [Schoenberg] does." He returns to the twelve-tone original. "Ah," Tommasini sighs, and strikes another dissonant chord. To me it sounds like a cat walked over the keyboard.

But chances are that if I listened to twelve-tone music often enough, I'd stop hearing cacophony and begin to think of it as "delicate" and "emancipating," just like Tommasini. The appeal of music, Voss and Clarke showed, is partly universal, reflected in the style-independent recurrences they discovered. Other researchers found it is also partly due to exposure during our upbringing, which shapes reactions to consonant and dissonant chords.[34] But we also value novelty. And professionals who make a living selling new ideas embrace opportunities to break with the boredom of the familiar.

In science too, our perception of beauty and simplicity is partly universal and partly due to exposure during education. And just like in music, what we perceive as predictable and yet surprising in science depends on our familiarity with the medium; we increase our tolerance for novelty through work.

Indeed, the more I read about the multiverse, the more interesting it becomes. I can see that it is an astonishingly simple and yet far-reaching shift in how we perceive our own relevance in the world (or lack thereof). Maybe Tegmark is right and I merely have an emotional bias against what is just a logical conclusion. The multiverse is truly emancipated mathematics, allowing full life and richness and complexity.

It also doesn't hurt if a Nobel Prize winner puts his weight behind it.

IN BRIEF

- Theoretical physicists use simplicity, naturalness, and elegance as criteria to assess theories.
- With naturalness now contradicted by observation, many physicists think that the only alternative to "natural" laws is that we live in a multiverse.
- But both naturalness and the multiverse require a metatheory that quantifies the probability of us observing the world the way it is, which is in conflict with simplicity.
- It is unclear what problem naturalness or the multiverse is even trying to solve, since neither one is necessary to explain observations.
- The multiverse is not universally considered beautiful, demonstrating that perceptions of beauty can and do change; their popularity cannot be explained by beauty alone.

6

The Incomprehensible Comprehensibility of Quantum Mechanics

In which I ponder the difference between math and magic.

Everything Is Amazing and Nobody's Happy

Quantum mechanics is spectacularly successful. It explains the atomic world and the subatomic world with the highest precision. We've tested it upside-down and inside-out, and found nothing wrong with it. Quantum mechanics has been right, right, and right again. But despite this, or maybe because of this, nobody likes it. We've just gotten used to it.[1]

In a 2015 *Nature Physics* review, Sandu Popescu calls the axioms of quantum mechanics "very mathematical," "physically obscure," and "far less natural, intuitive and 'physical' than those of other theories."[2] He expresses a common sentiment. Seth Lloyd, renowned for his work on quantum computing, agrees that "quantum mechanics is just counterintuitive."[3] And Steven Weinberg, in his lectures on quantum mechanics, warns the reader that "the ideas of quantum mechanics present a profound departure from ordinary human intuition."[4]

It's not that quantum mechanics is technically difficult—it isn't. The mathematics of quantum mechanics uses equations for which we have simple solution techniques, in stark contrast to the equations of general relativity, which are hideously difficult to solve. No, it's not the difficulty—it's that quantum mechanics doesn't quite feel right. It's disturbing.

It starts with the wave function. The wave function is the piece of math that describes the system you are dealing with. It is often referred to as the state of the system, but—and here's where it gets icky—it can't itself be observed by any conceivable measurement. The wave function is merely an intermediary; from it we compute the probability for measuring certain observables.

This, however, means that after a measurement the wave function must be updated so that the measured state now has a probability of 1. This update—sometimes referred to as "collapse" or "reduction"—is instantaneous; it happens at the same moment for the entire wave function, regardless of how far the function was spread out. If it was spread between two islands, measuring the state at one end determines the probability on the other end.

This isn't a thought experiment; it was actually done.

In the summer of 2008, Anton Zeilinger's group gathered on the Canary Islands to break the world record for long-distance quantum effects.[5] On the island of La Palma, they used a laser to produce 19,917 photon pairs; in each pair the total polarization was zero, but the individual polarization of each photon was unknown. Zeilinger's people sent one photon from each pair to a receiver on the island of Tenerife, 144 kilometers away. The other photon circled for 6 kilometers in a coiled optical fiber on La Palma. Then the experimentalists measured the polarization at both ends.

Since we know the total polarization, measuring the polarization of one photon tells us something about the polarization of the other photon. Just how much it tells us depends on the angle between the directions in which polarizations are measured at the two detector sites (Figure 9). If we measure polarizations in the same direction at both sites, then the measurement of one photon's polarization will reveal the polarization of the other. If we measure the polarization in two orthogonal directions, then measuring one tells us nothing about the other. At angles between 0 and 90 degrees we learn a little, and

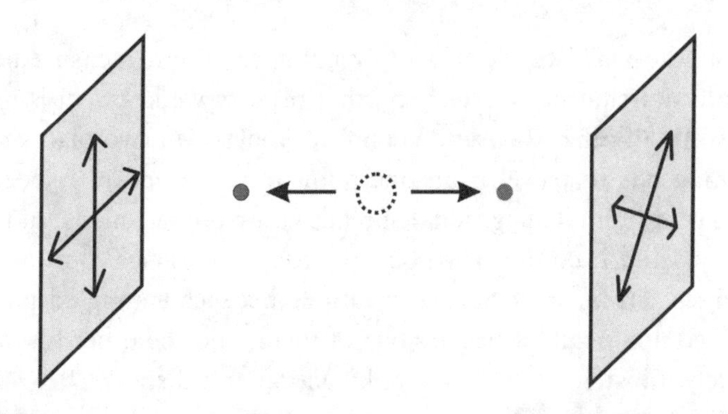

FIGURE 9. Sketch of experiment discussed in text. An initial state with total polarization zero (dotted circle) decays into two particles whose polarization must add up to zero. Then one measures the polarizations at two detectors (gray plates) in two directions with a relative angle.

the probability that both measurement outcomes agree quantifies just how much we learn.

We can calculate this probability without quantum mechanics by assuming that the particles already had a fixed polarization when they were created. But the result of this calculation doesn't agree with the measurement; it's just wrong. For some angles, the polarization measurements agree more often than they should. It seems that even though they are separated by 144 kilometers, the particles are linked more strongly than their common origin can explain. It is only when we do the calculation with quantum mechanics that the result comes out correctly. We must conclude that before measurement, the particles didn't have either one polarization or the other, but instead had both.

Zeilinger's experiment in the Canary Islands was neither the first nor the last to show that to correctly describe observations we must accept that unobserved particles can be in two different states at once. Though this experiment set a distance record (at the time), stacks of other experiments have demonstrated the same: quantum mechanics may be weird, but it's correct. Like it or not, we're stuck with it.

That the wave function simply collapses upon measurement is particularly irritating because no other process we know of is instantaneous. In all other theories, a connection between two places means something has to travel from one point to the other at a speed less than that of light. This gradual spread with passing time is known as "locality," and it conforms to our everyday experience. But quantum mechanics screws with our expectations because entangled particles are linked nonlocally. Measure one of them, and the other knows immediately. Einstein called it "spooky action at a distance."[6]

Then again, the nonlocality of quantum mechanics is subtle, because no information is exchanged between the entangled particles. Since the measurement outcomes for the polarizations cannot be predicted, there is no way to use the pairs to send messages from one end to the other. In quantum mechanics, as in non-quantum mechanics, information cannot be sent faster than the speed of light. Everything is mathematically consistent. It just…seems odd.

Another unappealing aspect of quantum mechanics is that by referring to measurements, the axioms assume the existence of macroscopic objects—detectors, computers, brains, et cetera—and this is a blow to reductionism. A fundamental theory should explain the emergence of the macroscopic world and not assume its presence in the axioms.

Quantum field theories inherit the problems from quantum mechanics. The standard model, therefore, has the same difficulty with explaining the macroscopic world.

To illustrate the problem with the emergence of a macroscopic, non-quantum world, Erwin Schrödinger in 1935 asked us to imagine a cat trapped in a box. The box has a mechanism that, when triggered by a decaying atom, can release a toxin and kill the cat. Atomic decay is a quantum process, so if the atom has a half-life of, say, 10 minutes, then there is a 50 percent chance the atom will decay within 10 minutes. But quantum mechanics tells us that before we've made the measurement, the atom neither has nor hasn't decayed. It's in a superposition of both options. So what about Schrödinger's cat? Is it both dead and alive, and only dies or survives the moment you open the box? It seems absurd.

It *is* absurd. There's a good reason we never witness quantum behavior in everyday life. For large objects—like cats or brains or computers—the quantum-typical properties fade away extremely quickly. Such objects are part of warm and wiggly environments, and the constant interactions scramble up quantum links between parts of the system. This scrambling—called decoherence—quickly converts quantum states to normal probability distributions, even in the absence of a measurement apparatus. Decoherence thus explains why we don't observe superpositions of large things. The cat isn't both dead and alive; there's just a 50 percent probability that it's dead.

But decoherence does not explain how after measurement the probability distribution updates to a distribution that has a probability of 1 for what we observed. So decoherence solves part of the riddle, but not all of it.

A Losing Game

I am still in Steven Weinberg's office in Austin. He just told me that his voice is getting hoarse and asks if we can cut the interview short. Ignoring his question, I say: "I wanted to ask you about your opinion on the foundations of quantum mechanics. You wrote in your book that it's hard to change anything about quantum mechanics without spoiling it altogether."

"Yes, that's right," Weinberg says. "But we don't have any really satisfactory theory of quantum mechanics."

He gives me a tired look, then grabs one of the books from his desk: "I can read to you what's in the second edition of my textbook on quantum mechanics. It's section 3.7. I wrote a rather negative section in the first edition, and then thinking about it more, I became even more negative. What I said here is:

> *My own conclusion is that today there is no interpretation of quantum mechanics that does not have serious flaws. This view is not universally shared. Indeed, many physicists*

> *are satisfied with their own interpretation of quantum mechanics. But different physicists are satisfied with different interpretations. In my view we ought to take seriously the possibility of finding some more satisfactory other theory to which quantum mechanics is only a good approximation.*

"And I feel that way. I have tried very hard to develop that more satisfactory other theory without any success.... It is very hard to do better than quantum mechanics. But quantum mechanics, although not inconsistent, has a number of features we find repulsive. It's not the fact that it has probabilities in it. It's what kind of probabilities there are."

He goes on to explain. "If you had a theory that said that, well, particles move around and there's a certain probability that it will go here or there or the other place, I could live with that. What I don't like about quantum mechanics is that it's a formalism for calculating probabilities that human beings get when they make certain interventions in nature that we call experiments. And a theory should not refer to human beings in its postulates. You would like to understand macroscopic things like experimental apparatuses and human beings in terms of the underlying theory. You don't want to see them brought in on the level of axioms of the theory."

ꝏꝏ

TO MAKE quantum mechanics more appealing, physicists have come up with various ways to reinterpret its mathematics. The two broad categories are either to take the wave function (usually denoted ψ, pronounced "psi") as a real thing (a "psi-ontic" approach) or, alternatively, to regard the wave function as a device that merely encodes knowledge about the world (a "psi-epistemic" approach).

Psi-epistemology doesn't so much answer questions as declare them meaningless. The best-known member of this class is the Copenhagen interpretation. It's the most commonly taught interpretation of quantum mechanics (and also the one I used in the previous section). According to the Copenhagen interpretation, quantum mechanics is a black box: we enter an experimental setup and push the math button, and out comes a probability. What a particle did before it was measured is a question you're not supposed to ask. In other words, "shut up and calculate," as David Mermin put it.[7] It's a pragmatic attitude, but the collapse is widely perceived as "an ugly scar" (Lev Vaidman) that makes the theory "kludgy" (Max Tegmark).[8]

A more modern psi-epistemic interpretation is QBism, where the "Q" stands for "quantum" and the "B" for "Bayesian inference," a method to calculate probabilities. In QBism, the wave function is a device that collects an observer's information about the real world, and it is updated when he or she makes a measurement. This interpretation emphasizes there can be multiple observers (people or machines) holding different information. The box is still black, but now everybody has their own. David Mermin calls it "by far the most interesting game in town," but for Sean Carroll it's a " 'denial' strategy."[9]

The psi-ontic interpretations, on the other hand, have the advantage of being conceptually closer to the pre-quantum theories we've come to like, but have the disadvantage that they force us to confront the other issues of quantum mechanics.

In pilot wave theory (or the de Broglie–Bohm theory), a nonlocal guiding field leads otherwise classical particles on paths. While it sounds entirely different than quantum mechanics, it is really the same theory just formulated and interpreted differently.[10] Pilot wave theory is currently unpopular because it hasn't been as generally formulated as the Copenhagen interpretation and can't be applied as flexibly. But because of its intuitive interpretation, Bell considered pilot-wave theory "natural and simple."[11] John Polkinghorne, on the other hand, thinks it has "an unattractively opportunistic air to it."[12]

The many-worlds (or many-histories) interpretation posits that the wave function never collapses. What happens instead is that the wave function splits or "branches" into parallel universes, one for each possible measurement outcome. In the many-worlds interpretation, there's no measurement problem, there's just the question of why we live in this particular universe. Stephen Weinberg finds all these universes "repulsive," but Max Tegmark finds the logic "beautiful" and believes that "the simplest and arguably most elegant theory involves parallel universes by default."[13]

Then there are spontaneous collapse models, in which the wave function doesn't first spread and then suddenly collapse, but constantly contracts a little, so that it never spreads very much to begin with. This isn't so much a new interpretation as an amendment to quantum mechanics that adds an explicit process for the collapse. Adrian Kent, who thinks that in quantum mechanics "elegance seems to be a surprisingly strong indicator of physical relevance," finds collapse "a little ad hoc and utilitarian" but still "considerably less ugly than the de Broglie–Bohm theories."[14]

And these are just the main interpretations of quantum mechanics. There are more, and each of them can be in several different states at the same time.

ꝏꝏ

STEVEN WEINBERG puts down his book on quantum mechanics. He looks at me and I try to read his expression, but I can't decide whether he's more bemused or annoyed that I'm still here. That raised eyebrow I noticed earlier, I now see, seems permanently stuck in the position.

Weinberg's *Lectures on Quantum Mechanics* was a latecomer among his textbooks, not appearing until 2012. Since then he has also published several papers about how to probe or better understand the foundations of quantum mechanics. It is clearly a topic that has been on his mind lately. I wonder why he began pursuing this line of research. What makes this a good problem to think about, given all the possible problems he could be thinking about?

"What is it that you don't like about decoherence that leaves you with a probability distribution?" I ask.

"You can very well understand quantum mechanics in terms of an interaction of the system you're studying with an external environment which includes an observer," he says.[15] "But this involves a quantum mechanical system interacting with a macroscopic system that produces the decoherence between different branches of the initial wave function. And where does that come from? That should be described also quantum mechanically. And, strictly speaking, within quantum mechanics itself there is no decoherence."

He coughs again. "Now, there is an attempt to deal with this, which denies decoherence and contemplates treating human beings completely quantum mechanically just like everything else, and that's the many-histories approach. In the many-histories approach, if you start with a pure wave function, it's always a pure wave function.[16] But as time evolves, it has many terms in it, each one of which contains a description of observers, and the observers in each term think they're seeing something different—like one observer sees the spin [of a particle as] up and the other sees the spin down.

"And while you could live with the history of the universe splitting into two branches, in this many-histories approach you have an endless continual production of unimaginably large numbers of histories of the universe.

"Well," Weinberg concludes, "that might be the way things are, and I don't know anything logically inconsistent about it. But it's so repulsive to imagine this vast number of histories."

"What makes it repulsive?"

"I don't know. It just is. It's the uncontrolled number. Because it isn't something that just happens when some physicist gets a government grant and does an experiment, it's happening all the time. Every time two air molecules collide, or a photon from a star strikes the atmosphere—every time something happens, the universe's history is continuously multiplied.

"I wish I could prove that this continuous splitting of histories is impossible. I can't. But I find it repulsive. Obviously the people who originated it don't find it repulsive. But that's what it is. Different people have different interpretations of quantum mechanics, that are all satisfactory, but they're all different."

"And they all think each other's theory is repulsive," I remark.

"Right. And I've dealt with that," Weinberg says, and sighs. "Even the great philosophers in this field have differed," he continues. "Niels Bohr originally thought that a measurement involves a departure from quantum mechanics, that you can't account for measurement in terms of purely quantum mechanical concepts. Then other people said no, you can, but you have to give up the idea that you can say what's happening, you have to just say these are the rules for calculating the probability of what you get when you make a measurement.... And then other people say quantum mechanics is perfectly good, it's just that you get infinite numbers of histories. That's the [many-worlds] approach....

"The difficulty, of course, is that you don't have to settle these issues," Weinberg continues. "I've had a whole career without knowing what quantum mechanics is. I tell this story in one of the books that my colleague Philip Candelas was referring to a graduate student whose career essentially disintegrated, and I asked what went wrong and he said, 'He tried to understand quantum mechanics.' He could have had a perfectly good career without it. But getting into the fundamentals of quantum mechanics is a losing game."

(If you quote this, you can be the first person to quote someone quoting someone quoting himself quoting someone.)

"Do you know the book *Beauty and Revolution in Science*, by a philosopher of science with the name McAllister?" I ask.

"I'm not familiar with it."

"He is trying to expand on Kuhn's concept of revolutions in science. McAllister's argument is that every revolution in science necessitates overthrowing the concepts of beauty that scientists have developed."

"Kuhn is much more radical," Weinberg says. "Obviously a revolution calls for an overthrow of something. Kuhn pictured this as an overthrow of everything, so that one generation can't understand the physics of an earlier generation. I think that's just wrong. But obviously if you have a scientific revolution you're overthrowing something."

He pauses, then adds: "I think the idea that this might be aesthetic judgements is not bad. For instance, the Copernican revolution occurred because Copernicus thought that the heliocentric system was much more attractive than the Ptolemaic system. It wasn't because of any data. Obviously that was an aesthetic judgment that differed from the aesthetic judgments before. And I think the Newtonian revolution may have occurred because Newton did not find force acting at a distance ugly, whereas Descartes did. So Descartes tried to make a very ugly picture of the solar system where everything was the result of direct pushing and pulling. And Newton was content with a force going like the inverse square acting at the distance. It was a change in aesthetic. Or you might say a change in philosophical preconception. Which might not be very different." He goes silent for a moment, then murmurs, almost to himself, "Yes, that's an interesting idea. I should look at this book."

"But if that is so," I say, going back to McAllister's point, "that in a revolution one needs to overthrow the conception of beauty in theory development, then what good is it to use the beauty concepts from the past?"

"Well, beauty is just a means to an end of getting successful theories," Weinberg says. "When your conception of beauty changes, the theories may continue to be correct." He offers an example: "I suppose that Maxwell may have felt that a really satisfying theory of electromagnetism would be one that involved stresses in a medium that was undergoing vibrations, and that would explain the oscillations of electric and magnetic fields observed in a ray of light. Through the work of Maxwell and other people, including Heaviside, we came to think of electric and magnetic fields as simply permeating empty space, and the oscillations are just oscillations in the fields themselves, not in an underlying medium. But the equations that Maxwell developed are still good. Maxwell's theory survives, even though his conception of why it should be true has changed."

He continues: "What changes very often is not physical theories, but our conception of what they mean, why they should be true.

So I don't think you overthrow everything, although you may overthrow previous aesthetic judgements. And what survives are the theories that the previous aesthetic judgements gave rise to. If they are successful—which they may not be."

Then he stands up and walks out.

Quantum Mechanics Is Magic

Not only is quantum mechanics itself weird, the research area is too. In particle physics we have theory, experiment, and, in the middle between them, phenomenology. Phenomenologists are the ones who (like Gordy Kane) coax predictions out of theories, usually by simplifying the math and figuring out what can be measured, to which precision, and how (and, not rarely, also by whom).

In other areas of physics, researchers don't assort into these three categories as clearly as they do in particle physics. But in all areas we have phenomenologists. Even in quantum gravity, where we don't have experiments, we have phenomenologists. Not so in quantum mechanics. In quantum mechanics, there's experiment on one side, and plenty of that. On the other side, there's much fuss about interpretation. But the middle between them is pretty much empty.

After looking into all these different interpretations and trying to assess their degrees of ugliness, I decide to speak to someone from the unfussy side, someone who handles quantum stuff in daily life. My choice is Chad Orzel.

Chad is a professor of physics at Union College in Schenectady, New York. He is better known as the person who taught quantum physics to his dog and wrote a book about it.[17] Chad also writes a popular science blog, *Uncertain Principles*, dedicated to demystifying quantum mechanics. I place a video call to him to ask what he thinks about all the quantum mechanics interpretations.

"Chad," I begin, "remind me what you do for a living."

"My background in physics is laser cooling and cold atom physics," Chad tells me. After obtaining his PhD, Chad worked on

Bose-Einstein condensates, clouds of atoms cooled to such low temperatures that quantum effects become strong.

"What I did for my PhD was looking at collisions of ultra-cold xenon atoms," Chad says. "They have relative velocities in the centimeter- or millimeter-per-second range, and at that speed they are moving so slowly that you start seeing quantum effects of the collisions.

"Xenon has a lot of isotopes," Chad explains.* "Some are composite bosons and some are composite fermions. And if you polarize them and they're fermions, they should be forbidden to collide because that would be two symmetric states and that's blocked."

This blockage is an example of fermions' extreme individuality, which we discussed in Chapter 1. You just can't force two fermions to do the same thing at the same place.

Chad continues: "So we're assembling a cloud of xenon atoms, and if they collide, they exchange a lot of energy and an ion comes out. We just count the ions for the case when the atoms are polarized and for the case when they are not, and this tells us how many of them collide. It's a very clean signal. And in the collision rate we can see the difference: the bosons happily collide and the fermions don't. This is a pure quantum effect."

"What do the atoms do if they don't collide?" I ask.

"They just pass right by each other," Chad says, and shrugs. "Somebody asked this in my thesis defense: 'What happens if you line these atoms up—how do they not collide?' And I said jokingly, 'Quantum mechanics is magic.' The more serious answer is that you should not think of them as little billiard balls that you can line up perfectly; you should think of them as big fuzzy things that pass through each other. I turned to my PhD supervisor, who had just won a Nobel Prize, and I asked, 'Do you agree? Does that work?' and he said, 'Yes, quantum mechanics is magic.' "[18]

ꝏꝏ

* The atomic nucleus of a chemical element has a fixed number of protons but can have various numbers of neutrons. These different variants of the same element are called isotopes.

IT MAKES sense, intuitively, that our intuition fails in the quantum world. We don't experience quantum effects in daily life—they are much too feeble and fragile. Indeed, it would be surprising if quantum physics were intuitive, because we never had a chance to get accustomed to it.

Being unintuitive therefore shouldn't be held against a theory. But like lack of aesthetic appeal, it is a hurdle to progress. And maybe, I think, this one isn't a hurdle we can overcome. Maybe we're stuck in the foundations of physics because we've reached the limits of what humans can comprehend. Maybe it's time to pass the torch.

ꝏꝏ

ADAM WORKS on microbial growth experiments. Adam formulates hypotheses and devises research strategies. Adam sits in the lab and handles incubators and centrifuges. But Adam isn't a "he." Adam is an "it." It's a robot designed by Ross King's team at Aberystwyth University in Wales. Adam has successfully identified yeast genes responsible for coding certain enzymes.[19]

In physics too the machines are marching in. Researchers at the Creative Machines Lab at Cornell University in Ithaca, New York, have coded software that, fed with raw data, extracts the equations governing the motion of systems such as the chaotic double pendulum. It took the computer thirty hours to re-derive laws of nature that humans struggled for centuries to find.[20]

In a recent work on quantum mechanics, Anton Zeilinger's group used software—dubbed "Melvin"—to devise experiments that the humans then performed.[21] Mario Krenn, the doctoral student who had the idea of automating the experimental design, is pleased with the results but says he still finds it "quite difficult to understand intuitively what exactly is going on."[22]

And this is only the beginning. Finding patterns and organizing information are tasks that are central to science, and those are the exact tasks that artificial neural networks are built to excel at. Such computers, designed to mimic the function of natural brains, now analyze data sets that no human can comprehend and search for correlations using deep-learning algorithms. There is no doubt that technological progress is changing what we mean by "doing science."

I try to imagine the day when we'll just feed all cosmological data to an artificial intelligence (AI). We now wonder what dark matter and dark energy are, but this question might not even make sense to the AI. It will just make predictions. We will test them. And if the AI is consistently right, then we'll know it's succeeded at finding and extrapolating the right patterns. That thing, then, will be our new concordance model. We put in a question, out comes an answer—and that's it.

If you're not a physicist, that might not be so different from reading about predictions made by a community of physicists using incomprehensible math and cryptic terminology. It's just another black box. You might even trust the AI more than us.

But making predictions and using them to develop applications has always been only one side of science. The other side is understanding. We don't just want answers, we want explanations for the answers. Eventually we'll reach the limits of our mental capacity, and after that the best we can do is hand over questions to more sophisticated thinking apparatuses. But I believe it's too early to give up understanding our theories.

"When young people join my group," Anton Zeilinger says, "you can see them tapping around in the dark and not finding their way intuitively. But then after some time, two or three months, they get in step and they get this intuitive understanding of quantum mechanics, and it's actually quite interesting to observe. It's like learning to ride a bike."[23]

And intuition comes with exposure. You can get exposure to quantum mechanics—entirely without equations—in the video game Quantum Moves.[24] In this game, designed by physicists at Aarhus University in Denmark, players earn points when they find efficient solutions for quantum problems, such as moving atoms from one potential dip to another. The simulated atoms obey the laws of quantum

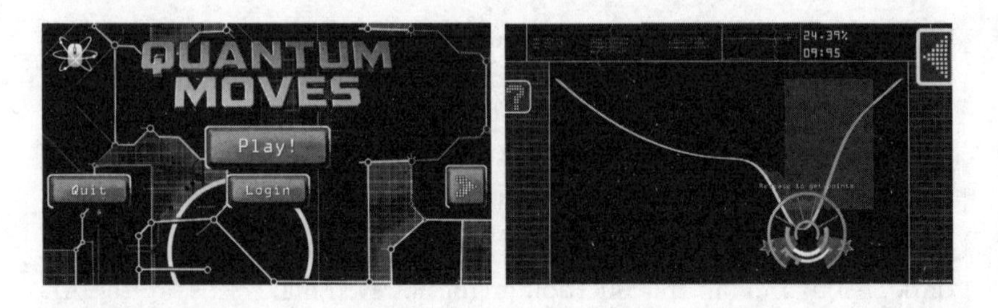

FIGURE 10. Screenshots of the video game Quantum Moves.

mechanics. They appear not like little balls but like a weird fluid that is subject to the uncertainty principle and can tunnel from one place to another. It takes some getting used to. But to the researchers' astonishment, the best solution they crowd-sourced from the players' strategies was more efficient than that found by a computer algorithm.[25] When it comes to quantum intuition, it seems, humans beat AI. At least for now.

So maybe, I think, we should just stop telling each other quantum mechanics is strange and get used to it. It's advanced technology, all right, but still distinguishable from magic.

CHAD'S JOKE that quantum mechanics is magic, accompanied by a shrug, makes me think he doesn't care much about interpreting the math. But it's hard not to interpret it. If you use a mathematical formalism frequently, you get a sense for what happens during the calculation. You don't just look at the result; you also see how you got there. And, as human interests go, we deal better with abstractions if they come with a story.

"Do you have a favorite interpretation of quantum mechanics?" I ask Chad.

"I'm temperamentally a shut-up-and-calculate person," he says. "It always seems to me that if you can't think of an experiment that you could do that gives different results for the different cases, it's kind of pointless. It's interesting to talk about what happens to all the weird stuff on the way to ordinary reality. But my understanding of the state of the field is that right now nobody can tell you an experiment that you could do that would give you a different result for, say, many-worlds versus [pilot wave]. And in the absence of this it's an aesthetic choice.

"But I do think it's valuable to have people pursuing all the different interpretations because it colors the questions that you think might be worth asking. And while you can explain all the experiments using any interpretation you like, some types of experiments arise more naturally using certain types of interpretations."

As an example, he names experiments that track the average curves along which particles move when passing through a double slit, curves that make sense in a pilot-wave interpretation but are meaningless if you think the wave function merely collects an observer's information. On the other hand, experiments that copy and erase quantum states are easier to interpret in terms of information transfer.

"Why is there so much argument over the 'right' interpretation if it's not tied to experimental test?" I ask.

"As I understand it," Chad says, "there's a divide between the epistemological and the ontological camps. In the ontological camp the wave function is a real thing that exists and changes, and in the epistemological camp the wave function really just describes what we know—it's just quantifying our ignorance about the world. And you can put everybody on a continuum between these two interpretations.

"On the one side people find offensive that you have this discontinuous collapse. This seems very ugly if you believe in the epistemic approach. On the other hand, there's the classic Einstein is-the-moon-there-if-nobody-looks argument."

Einstein believed that quantum mechanics was incomplete. Objects, he thought, should have definite properties whether or not someone observes them. The argument that it's absurd to think the moon isn't there when nobody looks exemplifies his thinking.[26] Keep in mind, however, that we do not expect large objects to have quantum properties due to decoherence. Like Schrödinger's cat, Einstein's moon is an exaggeration meant to illustrate a problem rather than an actual problem.

Chad lays out the appeal of Einstein's moon argument: "Things should exist whether or not people have any information about them. People want there to be an underlying reality, and on the ontological side they find it ugly to say this thing isn't really there until you measure some property. So on either side people find something offensive on the other side."

"Where are you on the spectrum?"

"I see a bit of merit in both ways of looking at it," Chad says. "I mostly agree that what we're learning is information about the world, but I'm also comfortable believing that there's a state that this [information] is *about.* So I'm in the wishy-washy middle."

I say, "In particle physics, we have people picking on things they don't like because they see such shortcomings as guides to a better theory. Is it similar in quantum foundations?"

"My impression is that it isn't quite the same as in particle physics," Chad replies. "In particle physics there are some very specific quantitative problems that you can point to and that we just can't answer. Like dark energy. We can do this calculation for the vacuum energy and it comes out 120 orders of magnitude too big, and then you have to do something wacky to make that go away. In quantum foundations, we can all agree [what happens when you send an electron through a double slit]. The question is what you believe happens on the way *to* it.

"Everybody can use the existing [mathematical] formalism and do the calculation and get the right result to some ludicrous number of decimal places. So it's not a quantitative problem. It's much more philosophical than the problems in particle physics. They both have an aesthetic component. They both feel like it shouldn't be this way because it's mathematically ugly. But in quantum foundations the quantitative disagreement isn't there. People aren't happy with the weird things that we know need to be true and are trying to find ways around this.

"A lot of the philosophical stuff that you get with quantum is only one level removed from the really ridiculous philosophy stuff, from Eugene Wigner's question 'Why can we describe things with mathematics anyway?'[27] And if that's the way it's posed, you lose a lot of sleep wondering why the universe obeys simple, elegant mathematical laws if there's no reason why that should be. But you can sweep this all away and say, 'Look, we *have* simple, elegant laws and we can do calculations!' And I feel like this."

"But when you're trying to find a new theory, is this a useful stance?" I ask.

"Yes, that's the issue there," Chad says. "Maybe we need to think about these philosophical questions and things that aren't calculable. But maybe the math is just ugly and someone needs to grind through it."

IN BRIEF

- Quantum mechanics works great, but many physicists complain that it's unintuitive and ugly.
- Intuition can be built by experience, and quantum mechanics is a fairly young theory. Future generations may find it more intuitive.
- In the foundations of quantum mechanics too, it is unclear what is the actual problem that requires a solution.
- Maybe understanding quantum mechanics is just harder than we thought.

7

One to Rule Them All

In which I try to find out if anyone would care about the laws of nature if they weren't beautiful. I stop off in Arizona, where Frank Wilczek tells me his little Theory of Something, then I fly to Maui and listen to Garrett Lisi. I learn some ugly facts and count physicists.

Converging Lines

The last time we had a theory of everything was 2,500 years ago. The Greek philosopher Empedocles postulated the world is made of four elements: earth, water, air, and fire. Aristotle later added a fifth element, the heavenly element quintessence. Explaining everything was never so easy again.

In Aristotle's philosophy, each element combines two properties: fire is dry and hot, water wet and cold, earth dry and cold, and air wet and hot. Change comes about because (1) the elements strive toward their natural place—air rises, rocks fall, and so on—and because (2) they can swap one of their properties at a time so long as no contradictions result, so dry and hot fire can turn to dry and cold earth, wet and cold water can change to wet and hot air, and so on.

Postulating that rocks fall down because that's their natural inclination doesn't explain much, but it arguably was a simple theory and one that could be summed up in a satisfactorily symmetric diagram (Figure 11).

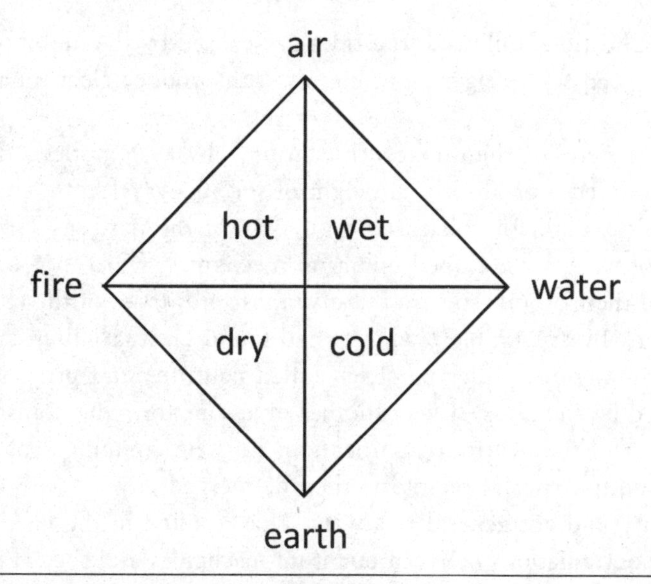

FIGURE 11. Schematic image of Aristotle's four elements and their relations, a theory dating back to the fourth century BCE. Similar classifications have been found in records of Eastern civilizations from around the same time.

Even in the fourth century BCE, however, it became apparent this was too simple. Alchemists started to isolate more and more substances, and a theory with only four elements couldn't explain this variety. It wasn't until the eighteenth century, however, that chemists understood that all substances are combinations of a relatively small number of "elements" (at the time, thought to be less than a hundred) that could not be further decomposed. The age of reductionism had begun.

Meanwhile, Newton figured that the falling of rocks and the movement of the planets have a common cause: gravity. Joule showed that heat is a type of energy, both of which were later found to originate in the motion of small things called atoms; there was a different type of atom for every chemical element. Maxwell combined electricity and magnetism into electromagnetism. And each time previously unrelated effects were explained by a common theory, new insights and applications followed: the tides are caused by the moon, energy can be used for refrigeration, circuits can produce electromagnetic radiation.

At the end of the nineteenth century, physicists noticed that atoms could emit or absorb only light of specific wavelengths, but they couldn't explain the observed regularity of the patterns. To make sense of it, they developed quantum mechanics, which not only explained these atomic spectra but also most properties of the chemical elements. By the 1930s, physicists had found out that all atoms have a core made of smaller particles called neutrons and protons, surrounded by electrons. It was another milestone for reductionism.

Next in the history of unification, Einstein combined space and time and got special relativity, then he merged gravity with special relativity and got general relativity. This resulted in the need to remove contradictions between quantum mechanics and special relativity, which led to the successful quantization of electrodynamics.

I think somewhere around here our theories were the simplest. But even back then physicists knew of radioactive decay, something that even quantized electrodynamics couldn't explain. A new, weak nuclear force was held responsible for the decays and was added to the theory. Then particle colliders reached energies high enough to probe the strong nuclear force, and physicists hit on the particle zoo.[1] This temporary growth in complexity was chopped down quickly by the theory of the strong nuclear force and the electroweak unification, which revealed that most of the zoo was composite, made up from merely twenty-four particles that were not further decomposable.

These twenty-four particles (together with the Higgs boson, added later, to make a total of twenty-five) are still elementary today, and the standard model plus general relativity still explain all observations. We've souped them up with dark matter and dark energy, but since we have no data about the dark stuff's microscopic structure, it isn't presently difficult to accommodate.

Unification, however, was so successful that physicists thought the logical next step would be a grand unified theory (GUT).

ꝏꝏ

WE CLASSIFY the symmetries of our theories with what mathematicians call a "group." The group collects all transformations that should not change the theory if the symmetry is respected. The symmetry group of the circle, for example, consists of all rotations around the center, and goes under the name U(1).

But in our discussion of symmetry so far, we have merely discussed the symmetries of the equations, of the laws of nature. What we observe, however, is not described by the equations themselves; it is instead described by a solution to the equations. And just because an equation has a symmetry doesn't mean its solutions have the same symmetry.

Think of a spinning top on a table (see Figure 12). Its environment is the same in all directions parallel to the table's surface, hence the equations of motion are rotationally symmetric around every axis vertical to the table. Once the top is given a spin, its motion is governed by the loss of angular momentum through friction. At first, the spinning does indeed respect the rotational symmetry. Eventually, though, the top tumbles and comes to a standstill. It will then point in one particular direction. We say the symmetry has been "broken."

Such spontaneously broken symmetries are commonplace in the fundamental laws of nature. As the example with the spinning top illustrates, whether a symmetry is respected by a system can depend on the energy in the system. The top, as long as it has sufficient kinetic energy, respects the symmetry. It is only once friction has carried away enough energy that the symmetry is broken.

The same is the case with fundamental symmetries. The energies that we normally deal with in daily life are determined by the temperature of the environment we find ourselves in. In particle physics terms, these everyday energies are tiny. Room temperature, for example, corresponds to about 1/40 of an eV, 14 orders of magnitude less than the energy the LHC pumps into proton collisions. At such low energy as room temperature, most fundamental symmetries are

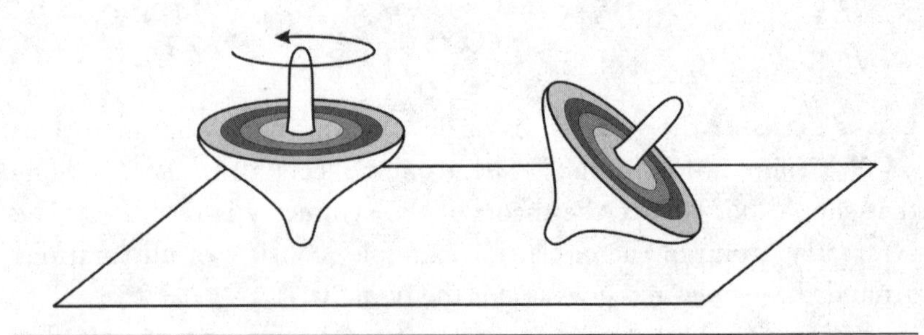

FIGURE 12. A spinning top, as long as it is in motion, is rotationally symmetric. As it loses energy through friction, the symmetry is broken.

broken. At high energies, however, they can become "un-broken" or "restored."

The symmetry of the electroweak interaction, for example, is restored at just about LHC energies, a signal of which is the production of the Higgs-boson.

ooooo

THE STANDARD model needs three different symmetry groups—U(1) and SU(2) for the electroweak interaction, and SU(3) for the strong nuclear force. These are small groups, as you can tell by the small numbers. But larger symmetry groups often contain several smaller groups, and therefore one large group whose symmetry is broken at high energy could give rise to the standard model at the energies we observe. In this picture, the grand unified symmetry is like an elephant, of which we have presently, at low energies, only an ear, a tail, and a leg. The full elephant would be restored only at the unification energy, estimated to be about 10^{16} GeV, or 15 orders of magnitude above LHC energies.

The first proposal for such a unified symmetry used the smallest group that contains the symmetry groups of the standard model, SU(5). Unified forces like this, however, generically enable new interactions that allow protons to decay. And if protons are unstable, so are all atomic nuclei. In such unified theories, the proton's lifetime could be as high as 10^{31} years, much exceeding the present age of the universe. But because of quantum mechanics, this would merely mean that the protons' *average* lifetime was that high. Once protons can decay at all, it can also happen quickly; it's just that quick decays are rare.

There are 10 protons in every water molecule, and 10^{25} water molecules in every liter of water. And so, instead of waiting 10^{31} years to see a single proton decay, we can monitor a large tank of water and wait for one of its protons to decay. Experiments like this have run since the mid-1980s, but to date nobody has seen a proton decay. The current observations (or lack thereof) imply that the average lifetime of protons is longer than 10^{33} years. With that, SU(5) unification is ruled out.

The next attempt at unification used a larger group, SO(10), in which the upper bound on the proton lifetime is higher. Since then, several other symmetry groups have been tried, some of which push the proton lifetime up to 10^{36} years, orders beyond even upcoming experiments.

Besides proton decay, grand unified theories also predict new particles because the large groups contain more than what's in the standard model. These new particles, as usual, are assumed to be too heavy to have been detected yet. And so theoretical physicists now have a selection of unified theories that are safe from being experimentally ruled out in the foreseeable future.

Grand unification alone, however, doesn't solve the problem with the Higgs mass. For that, physicists also supersymmetrize the grand unification. We know that supersymmetry—if it is a symmetry of nature—must be broken at energies above what we have tested so far, because we haven't yet seen susy particles. But we don't yet know at which energy the symmetry is restored—or if it happens at all. The argument that supersymmetry must render the Higgs mass natural means the energy at which susy is broken should have been accessible at the LHC.

Adding supersymmetry to grand unification doesn't only further enlarge the amount of symmetries; it has the added benefit that it tends to slightly increase the proton lifetime. Some variants of supersymmetric SU(5), for example, still linger at the edge of viability. The main reason to add supersymmetry, however, is another numerical coincidence that we discussed in Chapter 4: gauge coupling unification (Figure 8).

Grand unified theories, moreover, have a more stringent structure than the standard model, which adds to their appeal. For example, the electroweak theory is not a satisfactory unification because it still has two different groups—U(1) and SU(2)—and two coupling constants associated with them. These two constants are related by a parameter known as the "weak mixing angle," and in the standard model the value of this parameter must be determined by measurement. In most GUTs, however, the weak mixing angle is fixed by the group structure to 3/8 at GUT energies.[2] When extrapolated down to low energies, this is compatible with experimental data.

Many physicists don't think these numbers are a coincidence. I've been told so often that they must mean something, I sometimes also believe they must mean something. There are, however, a few "buts" that you should know of.

Most important, how well the gauge couplings converge to the same value depends on the energy at which supersymmetry is broken. If the energy is higher than 2 TeV or so, the convergence starts getting worse. The LHC is close to ruling out that the susy-breaking scale is below that, thereby ruining one of susy's greatest appeals. Moreover, if we want grand unification, there isn't any particular reason the couplings all have to meet at the same energy instead of two of them meeting first and then the third joining. It's just that this wouldn't be so pretty because an additional energy scale would have to be brought in.

Let me also mention that the convergence of the coupling constants is not exclusive to supersymmetry. It's a consequence of adding heavy particles, which begin to play a role at large energies. We can think up many other combinations of additional particles that will make these curves meet. In supersymmetry we're not free to choose the additional particles, and physicists think this rigidity speaks in favor of the theory. Furthermore, the meeting of the curves in susy was unexpected when it was first noted. And, as we have seen before, physicists pay more attention to unexpected finds.

So there are these buts. Supersymmetry, though, has something else going for it: some of the new supersymmetric particles would have the right properties to make up dark matter. They are produced abundantly in the early universe, are stable and hence stay around, and interact only weakly.

Supersymmetry therefore combines everything theoretical physicists have learned to cherish: symmetry, naturalness, unification, and unexpected insights. It's what biologists fittingly call a "superstimulus"—an artificial yet irresistible trigger.

"Supersymmetry provides a solution to all of these problems that is undeniably simpler, more elegant, and more beautiful than any other theory to have been proposed. If our world is supersymmetric, all of the puzzles pieces fit together nicely. The more we study supersymmetry, the more compelling the theory becomes," writes particle physicist Dan Hooper.[3] According to Michael Peskin, author of one of the most widely used textbooks on quantum field theory, susy is "the next step up toward the ultimate view of the world, where we make everything symmetric and beautiful."[4] David Gross calls it "beautiful and 'natural' and unique" and believes that "Einstein would, if he had studied [supersymmetry], have loved it."[5] And Frank Wilczek, though more cautious, trusts nature: "All these clues could be misleading, but that would be a really cruel joke by Mother Nature—and in really bad taste on her part."[6]

A Theory of Something

I'm in Tempe, Arizona, and there's a young man crossing the street against a red light. The shuttle bus driver hits the brakes and derails my train of thought with an explosion of insults. The man slowly clears the road, staring at his phone. Once done cursing, the driver tells me about the university's enlargement program, the increasing number of students that pop up on every corner, the new services catering to the students, and the new housing being constructed. I nod in approval of educated young Americans and the driver's efforts to not run them over.

Then he inquires about the purpose of my visit, and upon hearing I'm visiting the physics department, asks if I have something to do with CERN. Which I don't, really, but he just wants to know if I have ever been there. Yes, I've visited, and I've seen the tunnel and some magnets and the largest detector, ATLAS, as it was still under construction. Besides that, though, I mainly saw a lot of slides and drank too much coffee at a conference I have otherwise very little memory of. He hopes to one day visit CERN, he says, and he follows all the research and hopes we will discover something new. I feel guilty that we didn't, that we probably won't, and I'm afraid we'll leave him disappointed. I almost apologize.

The next morning I give myself two blisters by insisting on not wearing sneakers for once, and arrive just in time to find that the place I'm supposed to meet Frank Wilczek is closed. But I don't have to wait long until Frank shows up in a big hat, waving from down the street when he sees me. Not keen on further blisters, I opt for the place next door, a small bistro that offers breakfast.

Frank is a regular visitor at Arizona State University, and his present stay is hosted by the Origins Project, a transdisciplinary initiative with a strong outreach program dedicated to the exploration of foundational questions. Tonight he will be participating in a panel discussion moderated by Lawrence Krauss, director of the Origins Project. Frank gave the axion its name, has a fondness for the mantis shrimp, and has been practically drowned in awards.[7] In 2004, he won the Nobel Prize, together with David Gross and H. David Politzer, for the discovery that the strong nuclear force is "asymptotically free," meaning it becomes weaker at short distances. But given what I am concerned with, what makes him most interesting is his artsy side.

In his recent book *A Beautiful Question*, Frank asks whether "the world embodies beautiful ideas" and answers his own question with a resounding yes. But I already know that the world embodies beautiful ideas. I want to know whether the world embodies ugly ideas, and if so, whether we would continue to think of them as ugly.

I search through my backpack, sure that on the plane I noted down questions for Frank, but now I can't find them. Can't be so hard, I tell myself; he probably has a speech lined up anyway. Vaguely I ask, "What is the relevance of beauty in theoretical physics?" And off he goes.

"When we come to unfamiliar realms of reality—the subatomic, quantum realms—everyday intuition is unreliable," Frank begins. "And the idea of just gathering a lot of data, like Francis Bacon and Isaac Newton recommended, that's just not practical because experiments are so hard now. The way forward that has worked very well in the biggest part of the twentieth century was hoping that the equations would be very beautiful, that they would be symmetric and economical, and deriving consequences and then checking them. So that's what we do now—first we guess the equations and then check them."

"What makes a theory beautiful?" I ask. "You said it's economical and must have symmetry?"

"Symmetry is definitely part of it," Frank answers. "As you know, theories have gotten very sophisticated—there is local symmetry, space-time symmetry, anomalous symmetry, asymptotic symmetry—and all of these have proved to be very valuable concepts. But there is also a more primitive concept of beauty. You know it when you see it. It's when you're getting more out than you're putting in. It's when you introduce a concept to explain one thing and you find that it also explains something else. All these are somehow contributing to the sense that it's right.

"In most cases I think it's fair to call this an aspect of beauty," Frank continues, "but it's slightly different—it's economy and simplicity."

"Do you think that a better theory has to be simpler than the ones we have right now?"

"Well, we will see," Frank says. "Nature gets to do what nature does. I would certainly hope that it's simpler. But at the moment it's not clear. I guess the most serious candidate for a better theory—better in the sense of bringing together gravity with the other interactions in an intimate way and solving the [problem with quantum gravity]—is string theory. [But] to me it's not clear what the theory is. It's kind of a miasma of ideas that hasn't yet taken shape, and it's too early to say whether it's simple or not—or even if it's right or not. Right now it definitely doesn't appear simple."

"It would be easier to make a decision if we had some data," I point out.

"Yes, it would be much easier," Frank agrees. "But also, to have a theory of physics that doesn't tell you anything about the physical world is a very peculiar concept. So if you say we don't have data, that also means to me we don't have a physical theory. If we don't have data, what is it about?"

"It has a limit where you find gravity," I say, attempting a defense of string theory, "so at least you know that in some range of the parameter space it describes something that we see."

"If your standards are low enough, yes. But I don't think we should compromise on this idea of post-empirical physics. I think that's appalling, really appalling."

I hadn't known Frank followed the controversy about revising the scientific method to justify string theory. "I was about to ask you about this. I don't know how much you have followed the discussion..."

"Not in detail. That's all I had to hear, this idea of post-empirical physics."

"It goes back to this book by Richard Dawid, who started his career in physics but then turned to philosophy."

According to Richard, I explain, it is rational of string theorists to take into account all available information, including mathematical properties, to evaluate their theory.

"But what's the point of such a theory if it doesn't explain anything?" Frank asks.

"Richard is silent on this issue."

"Well, let me put it this way. If there was any bit of experimental evidence that was decisive and in favor of the theory, you wouldn't be hearing these arguments. You wouldn't. Nobody would care. It's just a fallback. It's giving up and declaring victory. I don't like that at all."

Yes, I think, the discussion about post-empirical arguments was brought on by lack of empirical arguments. But simply knowing that doesn't move us forward.

"But when it's so difficult and costly to make new experiments, it's unrealistic to expect that we will just stumble across some new data," I point out. "We need to decide where to look—and for that we need a theory. But this brings up the question, even in the absence of data, which theory do we work on? And that's why Richard says some theories are more plausible than others for various good reasons that theorists draw upon. For example, the absence of alternatives—the fewer alternatives that have been found, the more likely it is that the options which have been discovered are correct. At least according to Richard. But he doesn't touch the question of aesthetics. Which I find a huge omission—because physicists do use aesthetic criteria."

"Of course. But this discussion seems to me very abstracted and the most detached from reality. And there *are* exciting experiments; we don't need this. There are experiments looking for axions, the electric dipole moment, rare decays—and that's just particle physics. Then there is cosmology, gravity waves, odd objects in the sky. There could be all kinds of anomalies that could show up. We don't need theorists talking to themselves to guide experiments."

Yes, there are experiments, but for decades they've merely confirmed the already existing theories. For all I can tell, that theorists talk too much to themselves is both cause and effect of data starvation.

"The predictions for experiments in the last decades, they have not been terribly successful," I note.

"Well, they've found the Higgs particle," Frank says.

"Yes, but that was not beyond the physics of the standard-model," I say, leaving aside that the prediction dates back to the 1960s.

"Tangentially, the lightness of the Higgs is suggestive of supersymmetry," Frank says.

"Still heavy enough to get people worried susy might not be there."

"Yes," Frank agrees, "but it could have been a lot worse."

"They haven't found susy partners, though," I say. "Is this something that worries you?"

"I am starting to get worried, yes. I never thought it would be easy. There have been bounds from [the LEP experiments] and proton decay for a long time, and this indicated that a lot of the superpartners have to be heavy. But we have another good shot with the [LHC] energy upgrade. Hope springs eternal."

"Many people are worried now about susy not showing up at the LHC," I say, "because it means that whatever the underlying theory is, it is unnatural or fine-tuned. What do you think about this?"

"A lot of things are fine-tuned and we don't know why. I would definitely not believe in supersymmetry if it wasn't for the unification of gauge couplings, which I find very impressive. I can't believe that that's a coincidence. Unfortunately, this calculation doesn't give you precise information about the mass scale where susy should show up. But the chances of unification start getting worse when susy partners haven't shown up around 2 TeV. So for me that's the one reason to be optimistic."

"So you're not worried that the underlying theory might not be natural?"

"No, it could be next to minimal or some aspect of the symmetry breaking that we don't understand. These are complicated theories. I think we will learn from the experimental discoveries. Or we just learn to live with unnaturalness. The standard model itself is already very unnatural. The [parameter] that gives the electron its mass is already 10^{-6}. It just is," he says, and gives a shrug.

"So it doesn't really worry you."

"No. The case for susy would be even more compelling if, well, it had been discovered. More realistically, if it solved the hierarchy problem in a clear way. But for me by far the most powerful argument is the unification of coupling, and that has not gone away."

I ask, “So you would be fine with saying there are some underlying theories and these are the parameters and they just are what they are?”

“Well, eventually you would like to have better theory. Again—you don’t need to have a theory of everything to have a theory of something. But if a theory explains one thing well, that’s for me a very encouraging indication.”

“You probably know that Steven Weinberg has this speech about the horse breeder,” I say.

“The what? Horse breeder? No.”

“The horse breeder has seen a lot of horses, and now he looks at a horse and says ‘That’s a beautiful horse’ when, from experience, he knows that that’s the kind of horse that wins races.”

Frank says, “I think there are things about our sense of beauty that we can understand—it’s not totally mysterious. Certainly symmetry is part of it, and so is economy. It shouldn’t have loose ends. But I think trying to find an exact definition is too ambitious. Are you trying to find the exact circuits in your brain that make you think something is beautiful? For the most part people come to the same conclusion for what is beautiful.”

“Well, we’re all wired the same way, more or less,” I say. “And why should this sense of beauty be relevant for the laws of nature?”

“I think it’s the other way round,” Frank says. “Humans do better in life if they have an accurate model of nature, if their concepts fit the way things actually behave. So evolution rewards that kind of feeling that being correct gives you, and that’s the sense of beauty. It’s something we want to keep doing; it’s what we find attractive. So explanations that are successful become attractive. And over the centuries people have found patterns in what the ideas that work are. So we’ve learned to see them as beautiful.

“And now we also go through a process of training where [we] learn what has been successful. So we are trained to admire what is successful and what is beautiful. We also refine our sense of beauty as we gain experience, and we are encouraged by evolution to find the experience of learning rewarding.

"So that's my little theory of why the laws are beautiful. I don't think it's entirely inaccurate." Frank pauses, then adds: "Also, there is an anthropic argument. If the laws weren't beautiful, we wouldn't have found them."

"The question is, though, will we continue to find them?" I say. "We're looking at new things that we've never been exposed to before."

"I don't know," Frank says. "We'll find out."

"Do you know this book by McAllister?" I inquire.[8] "He is proposing an updated version of the Kuhnian idea that science proceeds through revolutions. According to McAllister, scientists don't throw out everything during a revolution; they only throw out their conception of beauty. So whenever there is a revolution in science, they have to come up with a new idea of beauty. He lists some examples for this: the steady state universe, quantum mechanics, et cetera.

"If that was true," I go on, "it would tell me that getting stuck on the ideas of beauty from the past is exactly the wrong thing to do."

"Yes, right," Frank says. "It's normally a good guiding principle. But occasionally you have to introduce something new. In each of these examples you find, though, that the new ideas are beautiful too."

"But people only found that new beauty after data forced them to look at it," I point out. "And I'm worried we might not be able to get there. Because we are stuck on this old idea of beauty that we use to construct theories and to propose experiments to test them."

"You might be right."

Strength in Numbers

I'm not very convinced by Frank Wilczek's little theory for why the laws are beautiful. If the perception of beauty was an evolutionary developed response to a successful theory, then why do so many theoretical physicists complain about the ugliness of the standard model, the most successful theory ever? And why do they find beauty in unsuccessful theories like SU(5) unification? I'll admit I find Frank's explanation pretty, but that doesn't mean it's correct.

And in any case, if we learn to find successful theories beautiful, that doesn't mean that we can use our sense of beauty to construct theories that are more successful; it just means that we'll construct more of the same theories. "If the laws weren't beautiful, we wouldn't have found them," Frank said. That's exactly what worries me. I'd rather have an ugly explanation than no explanation at all, but if he's correct, we might never find a more fundamental theory if it ain't beautiful enough.

"But," you might say, "there have always been scientists who got stuck on beautiful yet wrong theories. Scientists have always wrongly dissed colleagues who later turned out to be right. There has always been peer pressure, there has always been competition, there have always been confirmation bias, groupthink, wishful thinking, and professional hubris. But in the end it didn't matter. The good ones won, the bad ones lost. Truth prevailed, progress marched on. Science works, bitches. Why should it be any different now?"

Because science has changed, and continues to change.

MORE SCIENTISTS: The most obvious change is that there are more of us. The number of physics PhDs awarded in the United States has increased from about 20 per year in 1900 to 2,000 per year in 2012, growing by roughly a factor of 100.[9] The membership of the American Physical Society tell a similar story: it has increased from 1,200 in 1920 to 51,000 in 2016. For Germany, the stats are comparable: the German Physical Society today has about 60,000 members, up from 145 in 1900.[10] Judging by the number of authors in the physics literature, the global average has increased somewhat faster, by about a factor of 500 between 1920 and 2000.[11]

MORE PAPERS, MORE SPECIALIZATION: And more scientists produce more papers. In physics, the annual growth in paper output is about 3.8–4.0 percent per year since 1970, which corresponds to a doubling time of roughly eighteen years.[12] This makes physics a *slowly* growing area of science—a signal of the field's maturity.

As the body of physics literature has grown, it has fallen apart into separate subfields (a recent study identified ten major ones), each of which mostly references within itself.[13] Of these subfields, the most self-referential ones are nuclear physics and the physics of elementary particles and fields. Most of this book's topics fall into the latter area.

The specialization that this self-referential literature indicates improves efficiency but can hinder progress.[14] In a 2013 article published in *Science*, a group of researchers from the United States quantified the likelihood of topic combinations by looking at citation lists and studied the cross-correlation with the probability of the paper becoming a "hit" (by which they mean the upper 5th percentile of citation scores).[15] They found that quoting unlikely combinations in reference lists is positively correlated with a paper's later impact. They also note, however, that the fraction of papers with such "unconventional" combinations decreased from 3.54 percent in the 1980s to 2.67 percent in the 1990s, "indicating a persistent and prominent tendency for high conventionality."

MORE COLLABORATION: The increase in the total number of papers largely tracks the increase in authors. Remarkably, however, the number of papers per individual author has sharply increased in recent times, from about 0.8 per author per year in the early 1990s to more than twice as many in 2010.[16] That's because physicists coauthor more than ever before. The average number of authors per paper has risen from 2.5 in the early 1980s to more than double that figure today. At the same time, the fraction of single-authored papers has declined from more than 30 percent to about 11 percent.[17]

LESS TIME: Division of labor hasn't yet arrived in academia. While scientists specialize in research topics, they are expected to be all-rounders in terms of tasks: they have to teach, mentor, head labs, lead groups, sit on countless committees, speak at conferences, organize conferences, and—most important—bring in grants to keep the wheels turning. And all that while doing research and producing papers. A *Nature* survey in 2016 found that, on average, academic researchers spend only about 40 percent of their time on research.[18]

Hunting after money is especially time-consuming. A 2007 study found that university faculty in the United States and Europe spend another 40 percent of their working hours on applying for research funds.[19] In basic research, the process is particularly draining because of the now often mandatory prophecies about future impact. Predicting the fate of a research project in the foundations of physics is usually more difficult than doing the research itself.

It's therefore not surprising to anybody who has applied for research funds that interviews among academics in Great Britain and Australia revealed that lies and exaggerations have become routine in proposal writing. Participants in the study referred to their impact statements as "charades" or "made-up stories."[20]

LESS LONG-TERM FUNDING: The fraction of academics holding tenured faculty positions is on the decline, while an increasing percentage of researchers are employed on non-tenured and part-time contracts.[21] From 1974 to 2014 the fraction of full-time tenured faculty in the United States decreased from 29 percent to 21.5 percent. At the same time, the share of part-time faculty increased from 24 percent to more than 40 percent. Surveys by the American Association of University Professors reveal that the lack of continuous support discourages long-term commitment and risk-taking when choosing research topics.[22]

The situation in Germany is similar. In 2005, 50 percent of full-time working academics were on short-term contracts. By 2015, that figure had increased to 58 percent.[23]

LESS HETEROGENEITY: Academics today must constantly prove their value by producing measurable output. This doesn't make much sense because in some research areas it might take centuries until benefits become apparent. But since something has to be measured, academics are assessed by their current impact on the field they work in. The presently used measures for scientific success therefore heavily rely on publication counts and citation rates, which mostly measure productivity and popularity. Studies suggest that this pressure to please and publish discourages innovation: it is easier to get recognition for and publish research on already known topics than to pursue new and unusual ideas.[24]

Another consequence of the attempt to measure research impact is that it washes out national, regional, and institutional differences because measures for scientific success are largely the same everywhere. This means that academics all over the globe now march to the same drum.

In summary, we have more people, better connected than ever, who face increasing pressure to produce in specialized subfields with less financial security over shorter periods. This has made scientific communities an ideal breeding ground for social phenomena.

And so here's my little theory of something: Scientists are human. Humans are influenced by the communities they are part of. Therefore, scientists are influenced by the communities they are part of. Okay, I'm not going to win a Nobel Prize for this. But it leads me to conjecture that the laws of nature are beautiful because physicists constantly tell each other those laws are beautiful.

To offer you a different perspective, my mom likes to say that "symmetry is the art of the dumb." So what if I told you that a truly beautiful fundamental theory would be highly chaotic, not symmetric? Doesn't sound convincing? It'll get more convincing every time you hear it: research shows we consider a statement more likely to be true the more often we hear of it. It's called "attentional bias" or "mere exposure effect," and interestingly (or depressingly, depending on how you see it), this is the case even if a statement is repeated by the same person.[25] Chaos, really, is much more beautiful than rigid symmetries. Warming up to it yet?

But physicists presently don't reflect on the influence that belief-sharing has on their opinions. Worse, they sometimes use appeals to popularity instead of scientific arguments to support their convictions, like Giudice, who referred to the "collective motion" underlying the naturalness-trend. Or Leonard Susskind, who claimed in a 2015 interview that "almost all working high-energy theoretical physicists are convinced some sort of extra dimensions are needed to explain the complexity of elementary particles."[26] Or string theorist Michael Duff, who explained, "Rest assured that, if anyone found another more promising tree [than string theory], the 1,500 [string theorists] would start barking up that one." Not only do these scientists believe that if many people work on an idea, there must be something to it, but they also believe it's a good argument to make publicly.

Supersymmetry has particularly benefitted from social feedback. We have heard this from Joseph Lykken and Maria Spiropulu, according to whom it "is not an exaggeration to say that most of the world's particle physicists believe that supersymmetry *must* be true."[27] Or Dan Hooper: "The amount of time and money devoted to the pursuit of supersymmetry is staggering. It is hard to find a particle physicist who has not worked on this theory at some time in his or her career.... All over the world, thousands of scientists have been imagining a beautifully supersymmetric universe."

The Alternatives

But not everyone is a fan of supersymmetry.

One alternative that has drawn attention among the more mathematically inclined is that of Fields Medal winner Alain Connes, who reasons that supersymmetry "is a beautiful dream but it is too early to believe that this is the truth."[28] Connes has his own unified theory, and while it has gathered some followers, it hasn't caught on. At least in its present form, Connes's idea is far from aesthetic, culminating in a 384 × 384 dimensional matrix representation that Connes himself referred to as "daunting and not very transparent." It doesn't help that the mathematics he uses is far removed from what physics students are presently taught.

The gist of Connes's idea is this. In ordinary quantum theory, not everything comes in discrete chunks. The lines seen in atomic spectra—historically the first evidence for quantization—are discrete, but the position of a particle, for example, isn't; it can take on any value. Connes gave quantum behavior to positions, but he did so indirectly, through the way space-time can vibrate. Not only does this sidestep some of the usual problems with quantizing gravity, but amazingly, he also discovered that this allowed him to include the gauge interactions of the standard model.

Connes's approach works because vibrational modes of any shape contain information about that shape, and the curved space-time of general relativity is no exception. If vibrations are converted to sound waves, for example, we can "hear the shape of a drum"—if not by ear, then at least by mathematically analyzing the sound.[29] We don't have to hear the vibrations of space-time for them to be useful. Indeed, the vibrations don't even have to take place. What's important is that they are an alternative way to describe space-time.

The benefit of this alternative description, known as "spectral geometry," is that it can be made compatible with quantum field theory and generalized to other mathematical spaces—some of which contain the symmetry groups of the standard model. This is what Connes did. He found a suitable space and succeeded in getting back the standard model and general relativity in the cases in which we have tested them. He also made additional predictions.

In 2006, Connes and his collaborators predicted the Higgs mass to be at 170 GeV, way off the measurement of 125 GeV that would be made in 2012.[30] Even before the 2012 measurement, though, the

170 GeV figure was excluded by results that came out in 2008; upon this announcement, Connes commented, "My first reaction is of course a profound discouragement, mixed with an enhanced curiosity about what new physics will be discovered at the LHC."[31] After a few years, though, he revised the model, and now he argues that the previous prediction is no longer valid.[32] Either way, the approach has fallen out of favor.

There are also a few dissenters who hold on to the idea of "technicolor," according to which the particles currently thought of as elementary have a substructure made of "preons," which interact through a force similar to the strong force. This idea ran into conflict with data decades ago, but some variants have survived. Technicolor, however, is not currently particularly popular.

Since it's possible to combine fermions to make bosons, but not the other way round, there is the occasional attempt to build up all matter from fermions, such as spinor gravity or causal fermion systems.[33]

And then there is Garrett Lisi.

Off the Mainland

It's January and it's Maui, and I'm at the airport waiting for someone to pick me up. Around me, vacationers in colorful shirts vanish into hotel vans. I'm supposed to be visiting the Pacific Science Institute, courtesy of Garrett Lisi, the surfer dude with the theory of everything. But I have no address and my phone battery is dead again. My hair curls up in the humid Hawaiian air.

After thirty minutes—I just consider taking off my winter coat—a convertible stops at the curb. Garrett get outs and limps around the car, a big bandage on his left knee. "Welcome to Maui," he says, and places a lei around my neck. Then he squeezes my bag into the trunk.

On the way to the institute he tells me the bandage hides twenty-eight staples that hold together what's left of his knee. A paragliding accident—something about wind blowing one way instead of some other way—sent him smashing into volcanic rocks. The best part, he explains, is that it was all captured on video, since he was out there being filmed for a documentary.

His institute turns out to be a small house on a hillside with a lanai facing a large garden. There's a whiteboard in the living room and an illuminated red dragon in the kitchen. Photos of past visitors decorate the entry hall. It's after sunset, and Garrett's girlfriend, Crystal, hands me a flashlight so I can find the way to the little wooden guest cabin behind the house. She warns me about the centipedes.

A rooster wakes me early the next morning, and I hook onto the Wi-Fi. After a week of nonstop travel, my inbox is filling with urgent messages. There are two unhappy editors complaining about overdue reports, a journalist asking for comment, a student asking for advice. A form to be signed, a meeting to be rescheduled, two phone calls to be made, a conference invitation that needs to be politely declined. A collaborator returns the draft of a grant proposal for revision.

I remember reading biographies of last century's heroes, picturing theoretical physicists as people puffing pipes in leather armchairs while thinking big thoughts. I climb into a hammock, and there I hang until the surfer dude shows up.

∞∞

GARRETT DRAPES himself onto some patio furniture, carefully putting up his injured leg. He complains he hasn't been able to do any surfing for a week and apologizes for not being able to take me to the beach. That's fine by me, I assure him; the beach is not what I came for.

"Remember we had this conversation about Tegmark's mathematical universe, his claim that the universe is made of math? You said back then that, yes, the universe is made of math, but only the prettiest math."

"Yes, that's what I think," Garrett says. "But the question remaining is, which mathematics?"

"That's why I think Tegmark's idea is useless," I say. "It just moves the question from 'Which mathematics?' to 'Where are we in the mathematical multiverse?' And for all practical purposes that's the same question."

"Right," Garrett agrees. "It's interesting to talk about, but if you can't test it, it's not science."

"You could test it if you find something that can't be described by mathematics," I suggest.

"Ah! Well, that can't happen."

"Indeed," I say, "because we would never know that something can't be described by math, rather than that we just haven't been able to find out how. So why are you convinced that mathematics can describe everything?"

"All our successful theories are mathematical," Garrett says.

"Even the unsuccessful ones," I retort.

Undisturbed, Garrett launches into his story: "I didn't start out doing physics with this fanatical quest of finding something pretty to describe physics. I started out with the question 'What is an electron?' I really liked general relativity. But then I got to quantum field theory, and there you get this ugly description of electrons."

Garrett was looking for a description of fermions that is geometric in the same way that gravity is. In general relativity, what we call the gravitational force is a consequence of space-time curving around masses. Space-time's four-dimensional geometry can be fully grasped only by abstract math. But the conceptual similarity to rubber sheets makes general relativity tangible; it feels familiar, almost tactile. And, intriguingly, some of the elementary particles—the gauge bosons—can be described in a similar geometric way (though you still have to cope with these funny internal spaces). But this geometric approach doesn't work for the fermions.

"That's something that really bothered me as a graduate student," he says.

"Why did it bother you?"

"Because I figured the universe had to be described by one thing."

"Why?"

"Because it would seem that the universe has to be one consistent object," Garrett explains.

"Saying that you don't like how electrons are described isn't an inconsistency," I say.

"Right. But I *really* didn't like it."

Garrett calls the way that we currently describe fermions "not naturally geometric," emphasizing he uses the word "natural" differently than particle physicists do.

"So the fermions are not naturally geometric in the same way that gravity and the [gauge bosons] are. And this bothered me, a lot. But it didn't bother anybody else."

"What about supersymmetry?"

"I looked into supersymmetry. But the way supersymmetry is normally defined is very awkward. It's not natural, in the mathematical sense. And then, once you have this formalism describing bosons and fermions, it demands that there has to be a supersymmetric partner for every particle. And we haven't seen them. And we continue to not see them," he says, clearly pleased.

"So I wasn't a fan of susy. I wanted to find a natural description of fermions. This is what I set out looking for after my PhD. That's why I came out to Maui.

"At first I worked on the old Kaluza-Klein theory," Garrett tells me, "and I thought it was beautiful. . . . But I couldn't get the fermions to work out."[34]

Up to this point, Garrett's story is much like my own. I too started working on the old Kaluza-Klein theory, intrigued by its appeal but unhappy about the way it dealt with fermions. Unlike Garrett, however, I paid my rent from a three-year PhD scholarship, and my life aspirations were those of the middle-class, Middle European family I come from: a good job, a nice house, a child or two, a cozy retirement, and a tasteful urn. Maybe quantize gravity on the way. But moving to an island wasn't on the list.

When my Kaluza-Klein enthusiasm hadn't gotten anywhere after two years, my supervisor very vocally suggested I change topic and work instead on large extra dimensions—the Kaluza-Klein revival by Arkani-Hamed and collaborators, which was just about to bloom. I thought my supervisor had a point, and there my path diverges from Garrett's. I joined the particle physics community, financed myself on short-term grants and contracts, and at a regular rate dutifully produced papers about reasonably timely topics. Garrett took the road less traveled.

"After six years, I completely put all this Kaluza-Klein stuff aside," Garrett says. "I could do this because I have no academic inertia, because I worked by myself out here in Maui and had no students or grants."

Garrett started over and looked at the problem again from a different perspective. After years of hard work, he was rewarded with a breakthrough. All known particles, he found—including both fermions and bosons—could be described geometrically with one large symmetry. And unlike conventional grand unified theories, his symmetry also contains gravity.

ooo

FOR HIS theory, Garret uses the symmetry of the largest exceptional Lie group, E8. A "Lie group" (named after Sophus Lie [1843–1899], pronounced "Lee") is an especially pretty group because one can also do geometry in it, much like in the familiar space that we see around us. This is what Garrett was after.

There are infinitely many Lie groups. But by the late nineteenth century they had all been classified by Wilhelm Killing (1847–1923) and Élie Cartan (1869–1951). It turns out that most of the Lie groups fall into one of four families, each of which has an infinite number of members. The symmetry groups SU(2) and SU(3) of the standard model, for example, are Lie groups, and as the name indicates, they are quite similar in their structure. Indeed, there is a simple Lie group SU(*N*) for any *N* that is a positive integer. There are three other similar, infinite families of Lie groups, whose precise terminology need not concern us here. More important is that besides these four families, there are five "exceptional" Lie groups, named G2, F4, E6, E7, and the largest one, E8. And it can be proved that these are all the simple Lie groups there are, period.

To appreciate how bizarre this is, imagine you visit a website where you can order door signs with numbers: 1, 2, 3, 4, and so on, all the way up to infinity. Then you can also order an emu, an empty bottle, and the Eiffel Tower. That's how awkwardly the exceptional Lie groups sit beside the orderly infinite families.

ↀↀ

I WROTE IT up and put out a paper and it made a big splash," Garrett says, recalling the media attention that his paper attracted.[35]

Back then, he admits, his theory still had some shortcomings; for example the three generations of fermions didn't quite come out right. But this was in 2007. In the years since, Garret has solved some of the remaining puzzles. Still, his masterpiece is unfinished, not yet to his full satisfaction.

"But I do have a natural description for the fermions now," Garrett says. "So, in a sense I've accomplished what I set out after graduate school to do. And I found this big E8 Lie group that I wasn't expecting. I didn't go out looking for a pretty theory of everything—that would have been way too ambitious even for me."

"Well, it's a large group," I say. "Is it so surprising that you can fit a lot of stuff into this group? Too much stuff, even—my understanding is that you also have additional particles?"

"Yes, it's about twenty new particles," he admits, but hastens to add, "Which is not as many as in susy."

"I assume they're so massive, your additional particles, that we don't observe them?"

"Yes—the usual theorists' cop-out. [But] the action is uniquely beautiful.* It resembles the action for a minimal surface. In many

* The "action" is the math-thing that particle physicists use to define a theory.

ways it's the simplest possible action. It's hard to imagine that nature would want to modify that."

"How do you know what nature wants?"

"Well, that's the game. If you want to find a theory of everything, your aesthetic sense is pretty much all you have to work with."

"What makes it beautiful? You already said you have this geometric naturalness."

"Yes, it's natural because it all can be described with [geometry]. And you're using the largest simple exceptional Lie group. It's rich and yet simple. And it has these extensions deep into mathematics in different directions...that is really nice."

"You sound like a string theorist."

"I know! I know I sound like a string theorist! I share much of their motivations and desires. If I had gathered momentum with thousands of people for thirty years, working on E8 theory, and it failed, I would be in exactly the same position as they are now."

After the initial press attention faded away, Garrett Lisi and his E8 theory were quickly forgotten. Few in the physics community showed even the remotest interest in it.

"It hasn't attracted a lot of attention, has it?" I ask.

"No, not after it made the big news splash."

"Did something good come out of the hype?"

"My father no longer asks me when am I going to get a job," Garrett jokes. "Because I did very well in school and got a PhD and then...I went to Maui to become a surf bum, and my parents were like, 'What happened?' But I'm happy.

"When I'm working on physics," Garrett says, "I have to go to this place where I'm not thinking of anything else, only what's in front of me—the mathematics and the structures. When I'm doing this, I can't think about any other problems that are going on in my life. And in that sense it's an escape."

For a surf bum, he's surprisingly intellectual. No wonder the Internet loves him.

"Did you talk with Frank [Wilczek] about the bet that we made?" Garrett asks.

I had entirely forgotten about this. In July 2009, Garrett bet Frank $1,000 that no supersymmetric particles would be found in the next six years.

"But then the LHC had these hiccups with the magnets and so on," Garrett says. "So the bet came due last July, but that wasn't in the spirit of the bet because the data wasn't completely there yet, so we agreed to extend the bet by a year." It will be due in six months.

"Yes, there was this story that the LHC should find supersymmetry," I say. "Gordy Kane still thinks gluinos have to show up in run two."

"Ack," Garrett says. "The most egregious thing was his claim of predicting the mass of the Higgs, after all the rumors were already flying. And two days before the official announcement, he put out this paper. And then the announcement confirms the rumors and he calls it a prediction from string theory!"

Garrett isn't part of academia, and it shows. He doesn't worry about grants, or the future prospects for his students, or whether peer reviewers will like his papers. He just does his thing and speaks his mind. This doesn't sit well with many people.

In 2010, Garrett wrote an article for *Scientific American* about his E8 theory.[36] He calls it "an interesting experience" and remembers: "When it came out that the article would appear, Jacques Distler, this string theorist, got a bunch of people together, saying that they would boycott *SciAm* if they published my article. The editors considered this threat, and asked them to point out what in the article was incorrect. There was nothing incorrect in it. I spent a *lot* of time on it—there was absolutely nothing incorrect in it. Still, they held on to their threat. In the end, *Scientific American* decided to publish my article anyway. As far as I know, there weren't any repercussions."

"I'm shocked," I say, and mean it.

"String theorists are in a hard position," Garrett says, "because they've been promising a theory of everything for thirty years but it never panned out. They thought they'd just come up with this magic [extra-dimensional space] and get all the particles right before lunchtime. Now we have the full landscape—it's a total failure."

And yet Garrett's reliance on mathematics is so very similar to that of string theorists.

"Why are you so convinced that there is a theory of everything?" I ask.

"What we have now with the standard model, it's a mess," Garrett says. "We have the [mixing matrices], and the masses—we have all these parameters. I am of the opinion that all these pseudo-random parameters have an underlying explanation that will lead to a unified underlying theory."

"What's wrong with random parameters? Why does it have to be simple?"

"Well, as we've gone down in distance scales it's always become simpler," Garrett argues. "You start with chemistry, and the substances, they have all these scattered properties. But the underlying elements are pretty simple. And if you look at shorter distances, inside the atom, it gets even simpler. Now we have the standard model, which seems to be a complete set of particles and gauge bosons. And with what I have done, I think I have a good geometric description of the fermions. It looks like it's all one thing. To me, that's just extrapolating the path of science. Things seem to look simpler if we look at smaller scales."

"That's because you conveniently started with chemistry," I say. "If you start at a larger scale, the scale of, say, galaxies, and go down, it doesn't get simpler—it first gets more complicated, as with life crawling around on planets and all that. It's only past the level of biochemistry that it starts getting simpler again."

"Ah, it can't be like that," Garrett says. "We know the elementary particles can't be like planets. We know they are exactly identical."

"There's no such thing as 'exactly'—it's always to some limited precision," I point out. "But I don't mean that elementary particles are like planets," I explain. "Just that, whatever is the theory at short distances, it might not be simpler than what we have now. Simplicity doesn't always increase with resolution."

"Yes. It could be a mess," Garrett agrees. "Or it could be that there is some underlying framework that we'll never have experimental access to, and all we can see is this mess that sits on top of it. Xiao-Gang Wen, for example, he likes to say that the laws of nature are fundamentally ugly. But this is an idea that's abhorrent to me. I think that there is a simple, unified description that will explain all phenomena."

I make a note to contact Xiao-Gang. Just then a guy in shorts and a Hawaiian shirt appears on the lanai.

"This is Rob," Garrett says, introducing us. "He's here to fire up the BBQ."

Garrett is celebrating his birthday today, and he has invited a group of friends to roast dead animals.

FIGURE 13. The root diagram of E8, depicting Garrett Lisi's theory of everything. Each symbol is an elementary particle. Different symbols are different types of particles. The lines show which particles are connected by triality. No, I don't know what that means either, but it's pretty, isn't it?
Image courtesy of Garrett Lisi

"I thought the BBQ is in the evening," I say.

"His specialty is slow-cooked ribs," Garrett explains. "But you're vegetarian, so you won't appreciate this."

"So it will smell like ribs, like, the rest of the day?"

"Yes," Garrett says, and waves as if to make me go away. "I'm trying to chase you to the beach."

ꝏꝏ

I FINALLY GO to the beach. Sea turtles peek through the waves. The water is clear and the sand is white, its coarse grains oddly shaped. Garret's girlfriend, Crystal, tells me the grains are made by a little fish, the parrotfish, which munches coral and, when done digesting, returns the remainder to the ocean. The fish's native Hawaiian name, *uhu palukaluka*, translates to "loose bowels." Adult parrotfish each produce more than 800 pounds of sand a year, which adds up to tons and tons for the whole fish population.

I slather on sunscreen and dig my toes into the sand. There's a lesson to be learned here, I think: if you pile up enough of it, even shit can look beautiful.

IN BRIEF

- Theoretical physicists like the idea of a grand unified theory that merges the three interactions of the standard model into one. Several attempts at grand unification have run into conflict with experimental data, but the idea has remained popular.
- Many theoretical physicists believe that relying on beauty is justified by experience. But experience will not help us to find new laws of nature if those laws are beautiful in unfamiliar ways.

- Science has changed a lot in the last decades, but scientific communities have not adapted to this change.
- The current organization of academia encourages scientists to join already dominant research programs and discourages any critique of one's own research area.
- Having the support of a like-minded community affects how scientists assess the use and promise of theories they decide to work on.

8
Space, the Final Frontier

In which I try to understand a string theorist and almost succeed.

Just a Humble Physicist

It's January and it's Santa Barbara. I had meant to talk to Joseph Polchinski at the December meeting in Munich, but he canceled his attendance on short notice. As I visit Santa Barbara, Joe is on medical leave from the University of California here.

As part of my postdoctoral studies I spent a year in Santa Barbara, but the street address Joe gave me is not in a part of town I've ever been in. Out here, real estate is spacious. The bushes are neatly trimmed, the cars are shiny, and the grass is very green. I meander through the narrow roads along the foothills, far away from the familiar areas of affordable student housing. I finally find the house at the end of a cul-de-sac and pull up in front of the garage. It's early afternoon and it's sunny. A gardener comes by in a golf-cart-like vehicle. Palm trees sway in the wind.

My finger hesitates over the doorbell. I don't normally hunt down ill people at home. But Joe has been enthusiastic about meeting, saying that the topic of the Munich conference—which was "Why trust a theory?"—has been much on his mind. He has spent much of his life studying the mathematics of string theory. I am here to find out why we should trust the math.

ꝏꝏ

A BRIEF HISTORY of string theory goes like this: String theory was originally developed as a candidate to describe the strong nuclear interaction, but physicists quickly figured out that a different theory, quantum chromodynamics, was more suited to the task. However, they noticed the strings were exchanging a force that looked just like gravity, and strings were reborn as a contender for a theory of everything. All particles, so goes the idea, are made of strings in different configurations, but the string-substructure is so small we can't see it with presently accessible energies.

For consistency, string theorists had to posit that the strings inhabit a world that has not three dimensions of space but twenty-five (plus one dimension of time).[1] Since these additional dimensions haven't been seen, the theorists further assumed that the dimensions are of finite size or "compactified"—like a (higher-dimensional) sphere rather than an infinite plane. And since resolving short distances requires high energies, we wouldn't yet have noticed additional dimensions if they are small enough.

Next, string theorists discovered that supersymmetry was necessary to keep the vacuum of their theory from decaying. This brought down the total number of dimensions from twenty-five to nine (plus one time dimension), but the need for compactification remained. Since no supersymmetric particles had been observed, string theorists assumed that supersymmetry is broken at high energies, so superpartners, should they exist, wouldn't yet have been seen.

It didn't take long until it was noted that, even if broken at high energy, supersymmetry would lead to disagreement with experiment by enabling interactions that are normally forbidden in the standard model, interactions that hadn't been seen. And so was invented R-parity, a symmetry that, when combined with supersymmetry, simply forbids the unobserved interactions because they would conflict with the new symmetry postulate.

The problems didn't stop there. Until the late 1990s, string theorists had dealt with strings only in space-times that have a negative cosmological constant. When the cosmological constant was measured and turned out to be positive, theorists quickly had to invent a way to accommodate that. They developed a construction that works with the positive number, but string theory is still best understood for the case with a negative cosmological constant.[2] This case is what most string theorists still work on. It does not, however, describe our universe.

None of that would have mattered had the many amendments been successful in creating a unique theory of everything. Instead, physicists found that the theory allows a huge number of possible configurations, each originating in a different possibility of compactification and leading to a different theory in the low-energy limit. Since there are so many ways to build the theory—presently estimated at 10^{500}—the standard model is plausibly among them. But nobody has found it, and, given the huge number of possibilities, the odds are nobody ever will.

In reaction, most string theorists discarded the idea that their theory would uniquely determine the laws of nature and instead embraced the multiverse, in which all the possible laws of nature are real somewhere. They are now trying to construct a probability distribution for the multiverse according to which our universe would at least be likely.

Other string theorists left behind the foundations of physics entirely and tried to find applications elsewhere—for example, by using string theoretical techniques to understand collisions of large atomic nuclei (heavy ions). In such collisions (which are also part of the LHC's program), a plasma of quarks and gluons can be created for a short amount of time. The plasma's behavior is difficult to explain with the standard model, not because the standard model doesn't work but because nobody knows how to do the calculations. Nuclear physicists thus welcomed new methods from string theory.

Unfortunately, the string-theory-based predictions for the LHC didn't match the data, and string theorists quietly buried the attempt.[3] They now claim their methods are useful for understanding the behavior of certain "strange" metals, but even string theorist Joseph Conlon likened the use of string theory for the description of such materials to the use of a map of the Alps for traveling in the Himalayas.[4]

String theorists' continuous adaptation to conflicting evidence has become so entertaining that many departments of physics keep a few string theorists around because the public likes to hear about their heroic attempts to explain everything. Freeman Dyson's interpretation of the subject's popularity is that "string theory is attractive because it offers jobs. And why are so many jobs offered in string theory? Because string theory is cheap. If you are the chairperson of a physics department in a remote place without much money, you cannot afford to build a modern laboratory to do experimental physics, but you can afford to hire a couple of string theorists. So you offer a couple of jobs in string theory and you have a modern physics department."[5]

∞∞

AN ALTERNATIVE history of string theory goes like this: String theory was originally developed as a candidate to describe the strong nuclear interaction, but physicists quickly figured out that a different theory, quantum chromodynamics, was more suited to the task. However, they noticed the strings were exchanging a force that looked just like gravity, and strings were reborn as a contender for a theory of everything.

Remarkably, strings naturally fit together with supersymmetry, which had independently been discovered as the most general extension of space-time symmetries. Even more remarkably, while originally several different types of string theory were found, these different theories turned out to be related to each other by "duality transformations." Such duality transformations identify the objects described by one theory with the objects described by another theory, thereby revealing that both theories are alternative descriptions of what is really the same physics. This led string theorist Edward Witten to conjecture that there are infinitely many string theories, all related to each other and subsumed by a larger, unique theory, dubbed "M-theory."

And strings continued to surprise physicists. In the mid-1990s, they noted that rather than being a theory merely of strings, it also contained higher-dimensional membranes, "branes" for short. By using this new insight, string theorists could study black holes' higher-dimensional siblings and recover the already known laws for the thermodynamics of black holes. This unexpected match convinced even the skeptics that string theory must be a physically meaningful theory. While the physics of black holes still holds its mysteries, string theorists are getting closer to solving the remaining problems.

The physical intuition of string theorists also led to mathematical discoveries, especially concerning the geometric shapes of the compactified additional dimensions, the so-called Calabi-Yau manifolds.[6] Physicists found, for example, that pairs of geometrically very different Calabi-Yau manifolds are related by a mirror symmetry, an insight that had eluded mathematicians and has since given rise to much follow-up research. String theory also enabled the mathematician Richard Borcherds to prove the "monstrous moonshine conjecture," a relation between the largest known symmetry group—the monster group—and certain functions.[7] The intricate connection between string theory and the mathematics of the monster group has recently inspired others to explore the potential relevance of the monster group for understanding the quantum properties of space-time.

Research on string theory has also brought about the biggest breakthrough in foundational physics in the last decades, the "gauge-gravity duality." This duality too is an identification between the structures of two different theories that reveals that both theories are really describing the same physics. According to the gauge-gravity duality, some types of gravitational theories can equivalently be formulated as gauge theories, and vice versa.[8] This means in particular that physicists can use general relativity to do calculations in gauge theories that previously were mathematically intractable.

The implications of this duality are nothing but astonishing, since the dual theories do not work in the same number of dimensions: the space-time of the gauge theory has one dimension of space less than the space-time of the theory with gravity. This means that our universe—us included—can be mathematically squeezed into two dimensions of space. Like a hologram, the universe only appears three-dimensional but can really be encoded on a plane.

And it's not just a new worldview. String theorists have also applied the gauge-gravity duality to situations including the quark-gluon plasma and high-temperature superconductors, and while quantitative results have not yet been obtained, the qualitative results are promising.

Both stories are true. But it's more fun if you pick one and ignore the other.

ꝏꝏ

BESIDES WRITING one of the first textbooks on string theory, Joe played a major role in the theory's development, as he was the one to demonstrate that, rather than being solely about one-dimensional objects, string theory also contains higher-dimensional membranes. Just a few days ago, he put out a paper in which he laid out his thoughts about whether non-empirical criteria are useful to assess the promise of a theory, the theory in question being string theory.

I shake hands with his wife and son. After taking off my shoes, I tiptoe over the carpet and sink into the couch.

"What are your thoughts about this idea of Richard Dawid's non-empirical theory assessment?" I begin.

"I don't know what these words mean," Joe says. "I'm just a humble physicist trying to understand the world. But I think what he is saying is quite close to my own way of thinking. If I ask myself, 'What do I want to work on, based on all the evidence I have gathered in my life? What are the most promising directions? What are the most probable directions for success?' then I need to make an assessment like this. And I believe that there is positive evidence—I count six types—that string theory is the right direction.[9]

"Some of the things that Dawid talks about seem to coincide with the way that I think about the problem. At the same time, these words, 'non-empirical'...I don't think about them."

He looks at me expectantly.

"I believe what Dawid means is exactly what you said," I agree. "And we clearly take into account other facts than data—this has probably always been the case. But now the time lag between developing and testing a hypothesis has become very large, and this makes the assessment of theories by other means than data increasingly important."

"Yes," Joe says. "[Max] Planck figured this out over a hundred years ago, that we're twenty-five orders of magnitude—then—between where they could measure and where we would probably have to go. Still today there are fifteen orders of magnitude to go. There is a large hope that we could see something at lower energies—this is your subject, quantum gravity phenomenology: the attempt to look for all possible ways to see things that would be accessible at lower energies. It's something we'd all want. Unfortunately, everything we've seen so far is negative.

"I think," he continues, "you have tried very hard to separate between ideas that sound like good ideas and ideas that do not seem like good ideas. I really look at you as someone who personally has tried very hard. It's an important thing to do. It's really thankless, though, because the number of bad ideas increases much more rapidly than the number of good ideas. And it takes a lot more time sometimes to find out why something is wrong than to produce something which is wrong."

This is possibly the nicest way I've ever been told I'm stupid.

"The people looking for phenomenology," Joe goes on, "they all have the same problem: they have to reach across the remaining fifteen orders of magnitude. It's a very hard problem. For most of string phenomenology it's not so much that you have a theory but that you have a possible phenomenology that someday may be part of the theory. And it's not because people are doing the wrong thing but because deriving phenomenology is such a difficult problem. And so in every part of the subject we have to think in much longer time scales than we are used to.

"And it's all due to Planck. If he'd just have come up with a smaller number..."

ꝏꝏ

THE PLANCK energy is where we should start to notice the quantum fluctuations of space-time. It's at approximately 10^{18} GeV, gigantically large compared to the energies that we can reach with colliders (see Figure 14). The large gap between the presently accessible energies and the energies at which grand unification and quantum gravity should become relevant is often called "the desert" because, for all we know right now, it might be void of new phenomena.

If we wanted to directly reach Planckian energies, we'd need a particle collider about the size of the Milky Way. Or if we wanted to measure a quantum of the gravitational field—a graviton—the detector would have to be the size of Jupiter and located not just anywhere but in orbit around a potent source of gravitons, such as a neutron star. Clearly these aren't experiments we'll get funded anytime soon. Hence many physicists are pessimistic about the prospects of testing

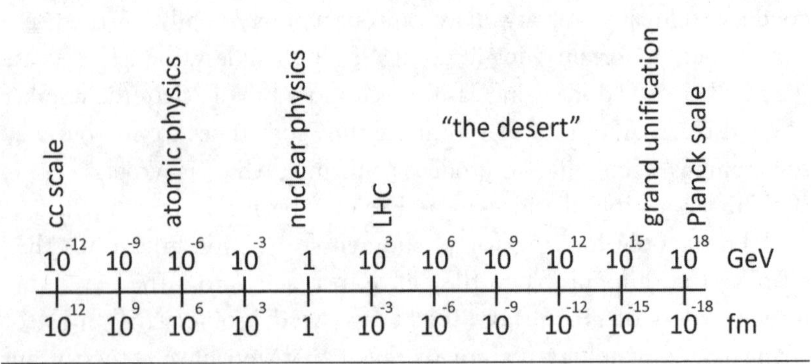

FIGURE 14. Energy scales. CC stands for cosmological constant.

quantum gravity, which leads to a philosophical conundrum: if we can't test it, is it science?

But few theoretical physicists worry about this conundrum, because this one is not only a question of aesthetics but also a question of consistency.

The standard model and general relativity combined lead to internal contradictions that, above the Planck energy, can't describe any observation. We therefore know that simply combining the two theories is wrong and there must be a better way to do it.

The origin of the contradiction is that general relativity is not a quantum theory but nevertheless must react to matter and radiation, which have quantum properties. According to the standard model, an electron, for example, can be in two places at once because it is described by a wave function. And according to general relativity, the mass of the electron curves space-time around it. But around which location? General relativity cannot answer this question, since its curvature doesn't have quantum properties and can't be in two places at once.

We can't measure this, because the gravitational pull of an electron is too weak, but that doesn't matter—a theory should be able to answer questions unambiguously regardless of whether it can be tested.

Such questions of consistency are rare and extremely powerful guides. The Higgs boson is an example of such a prediction of necessity. The standard model without the Higgs becomes internally inconsistent at the energy scales accessible at the LHC because the results of some calculations become incompatible with the probabilistic interpretation.[10] Therefore, we knew that something had to happen at the LHC.

Since no proof is ever better than its assumptions, it wouldn't have been possible to prove that something specific had to happen at the LHC. It could have been something besides the Higgs—the electroweak interaction could have turned out to become unexpectedly strong, for example. But we knew *something* had to happen, because the theories we had so far weren't consistent. If you want to stretch your brain, you could try to imagine nature showing an actual logical inconsistency or maybe adhering to a more complicated type of logic. But that too would mean "something new."

The expectation, however, that other new particles besides the Higgs should show up at the LHC was born not out of necessity but from the belief that nature tries to avoid fine-tuned parameters.

∞∞

THE HIGGS mass is a great surprise," Joe says, "because supersymmetry has, so far, not been found. And even if it is found now, the numbers are so pushed already, there is still a lot of fine-tuning. I don't know what to think of this. But I also don't have a better answer. Because the cosmological constant is such a big problem. And you need to do something about the Higgs mass.

"At first, we had two ideas—technicolor and supersymmetry—to solve the problem with the Higgs mass. With technicolor the solution was making the particles composite. Unfortunately, that idea seemed to rapidly become more complicated and less promising. Now, though, supersymmetry is in the same state. It used to be a beautiful thing, something that was very easy to root for, but now it's becoming more difficult. I still hope that we find it. And then maybe we can understand why it's realized at high energies in a more subtle way than we expected.

"I don't have a sense for why supersymmetry has not been seen and what it means for the future. Everybody is now excited because there's the 750 GeV bump.* You know how it is."

"What makes susy so beautiful?" I ask.

"I am always a little careful using words like 'beauty.' They're ill-defined," Joe says. "I wrote a review once about Dirac's work.[11] Dirac was very much motivated by beauty. In my review I said at the end, 'One can recognize beauty when one sees it, and here it is.' But I guess in some sense I avoid the word.

* This is the diphoton anomaly.

"I think I am less motivated by geometry than most people are. The ideas that are striking to me are the ones that connect things that were not obviously previously connected. Like we know the world has bosons and fermions and, extrapolating the way our field has moved forward in the past, it would be nice if there was some way that bosons and fermions would be connected.

"So, supersymmetry provided the kind of connections one was hoping to see. It provided the connection between fermions and bosons, and for why the Higgs wasn't heavy. It provided the potential to also do it for the cosmological constant. But it didn't work for the cosmological constant, and there aren't any good ideas now how to do it..."

He trails off. Finally he concludes, "Maybe all of this is just true at higher energies and not useful for what we observe now."

For a while he stares past me, out the window. Then he abruptly asks if I'd like a cup of coffee or something else to drink. I decline, and he grabs some notes.

Walls of Fire, Set in Stone

In preparation for the Munich conference, Joe compiled a list with mathematical evidence that speaks in favor of string theory. His list, I note, fits well with the aspects of beauty I already heard about.

String theory, Joe tells me, convinces him first and foremost because it succeeds in quantizing gravity, a problem to which not many solutions are known. Moreover, once you buy into the idea of strings, you do not have much freedom to construct the theory. These two reasons, I think, capture the appeal of "rigidity," which Nima Arkani-Hamed and Steven Weinberg also mentioned.

Joe then mentions that another feature in favor of string theory is that it is geometric—the aspect that was so important for Garrett Lisi—though Joe adds that for him, this "does not count so much."

The next two points on Joe's list are cases of what he calls "connections" and what philosopher Dawid referred to as "explanatory closure." They deliver the surprise necessary to make a theory elegant. The connections that Joe names are (1) the new insights revealed by the gauge-gravity duality ("We live in a hologram") and (2) string theory's contribution to black hole thermodynamics.

∞∞

BLACK HOLES form when a sufficiently large amount of matter collapses under the gravitational pull of its own mass. If the matter fails to build up sufficient internal pressure—for example, because a star has exhausted all its fuel—then it can continue to collapse until concentrated in a single point. Once the matter is concentrated enough, the gravitational pull at its surface becomes so strong that not even light can escape: a black hole has been created. The boundary of the trapping region is called the "event horizon." Light launched directly at the horizon will just about fail to escape, going around in a circle forever, and since nothing travels faster than light, nothing can escape from inside the black hole.

The horizon is not a physical boundary. It doesn't have substance, and its presence can be inferred only from a distance, not as you approach it. Indeed, you can cross the horizon without even noticing, provided the black hole is large enough. That's because in free fall we don't experience gravitational pull, only a change in the pull, known as tidal force. And the tidal force is inversely related to the mass of the black hole: the larger the black hole, the smaller the tidal force.

Indeed, if you fell into a supermassive black hole—like the one at the center of the Milky Way—the tidal force is so small that you wouldn't notice when you cross the horizon. Assuming you dove in headfirst, the tidal force would pull your head a little more than your feet, so you get stretched. At the horizon the stretch is minuscule. As you get closer to the black hole's center, the stretch starts becoming uncomfortable, but by then it's too late to turn around. The technical term for the cause of your death would be "spaghettification."

Black holes used to be speculative, but in the past twenty years astronomers have gathered overwhelming evidence for their existence, both for stellar-mass black holes (formed from burned-out, collapsed stars) and supermassive black holes (with masses from 1 million up to 100 billion times that of our Sun). Supermassive black holes are found in the center of most galaxies, though it is still unclear exactly how supermassive black holes grow to their size. The one in our own galaxy is called Sagittarius A* (pronounced "A-star").

The best observational evidence for black holes we presently have comes from the orbits of stars and gas in their vicinity, combined with the absence of radiation that should come from accreted matter when it hits a surface. The orbits tell us how much mass is squeezed into the observed region of space, and the lack of radiation tells us the object can't have a hard surface.

But black holes don't fascinate only experimentalists, they also fascinate theorists. What intrigues them most are the consequences of a calculation by Stephen Hawking. In 1974, Hawking demonstrated that even though nothing can escape the horizon, a black hole can still lose mass by radiating off particles. The particles of what is now called "Hawking radiation" are created by quantum fluctuations of matter fields in the vicinity of the horizon. The particles are produced in pairs from energy in the gravitational field. Every once in a while, one particle of the pair escapes, while the other one falls in, leading to a net loss of mass from the black hole. This radiation consists of all types of particles and is characterized by its temperature, which is inversely proportional to the mass of the black hole—meaning that smaller black holes are hotter, and a black hole heats up as it evaporates.

Let me emphasize that Hawking radiation is *not* caused by quantum effects of gravity, but rather is the product of quantum effects of matter in a curved non-quantum space-time. That is, it is calculated using only theories that are already well confirmed.

Why is the evaporation of black holes so fascinating for theorists? It's because Hawking radiation does not contain any information (besides the value of the temperature itself); it is entirely random. But in quantum theory, information cannot be destroyed. It can become so wildly mixed-up that it is in practice impossible to recover, but in principle quantum theory always preserves information.[12] If you burn a book, its information only seems lost; really, it just becomes transformed into smoke and ashes. Though the burned book is no longer useful to you, it's not in conflict with quantum theory. The only process we know of that really destroys information is black hole evaporation.

Hence the conundrum: we started trying to combine gravity with the quantum theory of matter, and found that the result isn't compatible with the quantum theory. Something has to give, but what? Most theoretical physicists—me included—think we need a theory of quantum gravity to resolve this issue.

Until 2012, many string theorists believed they had solved the information loss problem with the gauge-gravity duality. With this duality, whatever happens during black hole formation and evaporation can alternatively be described by a gauge theory. In the gauge theory, however, we know that the process is reversible, and so the black hole evaporation must be reversible as well. This doesn't explain just how the information comes out of the black hole, but it demonstrates that in string theory the problem is absent. Better still, using this method, string theorists can count how many ways there are to make a black hole—the black hole "microstates"—and the result fits perfectly with Hawking's calculation of the temperature.

Everything looked good for string theorists. But then something unexpected happened: A calculation by Joe Polchinski and collaborators from the University of California, Santa Barbara, proved that what they thought was the right explanation can't be right.[13]

Hawking radiation—the type that doesn't contain information—is compatible with general relativity, according to which a freely falling observer shouldn't notice crossing the event horizon. But Polchinski and collaborators demonstrated that if one forces information into the Hawking radiation, then the horizon must be surrounded by highly energetic particles that rapidly burn anything and anyone falling into a black hole—black holes would be surrounded by what they call a "firewall."

The firewall created a lose-lose situation for string theorists: destroy the information and ruin quantum mechanics, or release the information and ruin general relativity. But neither option is acceptable for a theory whose purpose was to combine quantum mechanics and general relativity.

In the four years after the firewall paper was published, it's been cited more than five hundred times, but no agreement what to do about it has been reached.

The temperature of solar-mass and supermassive black holes is so small it can't be measured—it is much below the already tiny temperature of the cosmic microwave background. The black holes we can observe at present gain more mass by swallowing their environment than they lose by Hawking radiation. There is thus no way to experimentally probe any of the attempts to understand black hole evaporation. It's a pure math problem with no risk of interference by data.

Math Versus Hope: A Case Study

The sixth and final reason on Joe's list of mathematical evidence in favor of string theory is the multiverse. Listing the multiverse as a desirable aspect of string theory did not come easily to him, however.

After obtaining his PhD, Joe tells me, he tried to explain the value of the cosmological constant—at the time thought to be zero—but couldn't find an explanation. Then Steven Weinberg proposed that the cosmological constant is a random parameter, in which case we can merely calculate what is the most probable value that we could observe.

"When Weinberg came up with his argument, I said no," Joe tells me. "I wanted to compute this number, I didn't want it to be a random number."

Weinberg didn't say where all the random values of the cosmological constant exist; he just assumed there must be a large number of universes. It was a rather vague idea at that point, really. But that would change, and Joe played a role in that.

He recounts: "[Raphael] Bousso and I then showed, to my great despair, that string theory seemed to provide exactly the kind of microscopic law that Weinberg needed." Math had revealed another connection and this one wasn't only unexpected but also unwelcome.

"I wanted this to go away, but it didn't go away," Joe says. "Even after people started working on this and started to study this, I wanted it to go away. I literally had to go to the psychiatrist over this. It made me so unhappy. I felt like it was taking away one of our last great clues as to the basic nature of fundamental physics, because things we had hoped to calculate now became random."

The cosmological constant was measured, and Weinberg's prediction was spot-on. The multiverse had demonstrated its use.

Joe recalls: "Sean Carroll reminded me several years later that I promised him he could have my office when the cosmological constant was found because I felt it would be the end of physics. For long years I felt that a large part of our way forward was blocked.

"To be honest," Joe adds, "I have a large tendency to be anxious that sometimes has made my life quite difficult. But with the multiverse, I finally came to the point where I thought I should probably see a doctor. It's true, I ended up going to the psychiatrist because of the multiverse," he says with a laugh.

But Joe slowly came to accept the new situation. He now thinks it's a plus that string theory provides a landscape of solutions which makes probabilistic predictions possible.

"Traveling back to the science story and throwing away this issue of what I wanted to be true," Joe goes on, "string theory provided this piece that Weinberg needed to complete his picture."

ᴑᴑᴑᴑ

FOR ME, Joe's story discloses a difficulty scientists rarely speak about: it can be hard to accept the truth, especially if it's ugly.

Finding beauty and meaning in the natural order of things is a human desire and one that scientists aren't immune to. The psychologist Irvin Yalom has identified meaninglessness as one of our four existential fears, and we work hard to avoid it.[14] Indeed, many cognitive shortcomings, such as our predisposition for wishful thinking (which psychologists prefer to call "motivated cognition"), are there to protect us from the harshness of reality.

But as a scientist, you have to let go of comforting delusions. This isn't always easy. What your equations reveal might not be what you hoped for—and the price can be steep.

Joe is one of the most intellectually honest people I know, always willing to go with an argument regardless of whether he likes where it takes him—as the examples with the black hole firewall and the multiverse demonstrate. This makes him an exceptionally clear thinker, though sometimes he doesn't like at all the conclusions logic has forced on him. And this is the very reason we use math in theoretical physics: if the math is done correctly, the conclusions are undeniable.

But physics isn't math. Even the best logical deduction still depends on the assumptions that we start from. In the case of string theory those are, among others, the symmetries of special relativity and the procedure of quantization. There's no way to prove these assumptions themselves. In the end, only experiment can decide which theory of nature is correct.

∞∞

JOE HAS come to the end of his notes. He says that for him, the biggest problem string theory presently faces is black hole firewalls. They "have taught us that we didn't know as much as we thought we did," he explains.

"How far, do you think, are we on the way to a theory of everything?" I ask.

"I hate the term 'theory of everything' because 'everything' is just ill-defined and presumptuous," Joe remarks as an aside. Then he says, "I think string theory is incomplete. It needs new ideas. These might come from loop quantum gravity. If that is so, there's going to be a fusion of direction. . . . This may be the direction we need. But string theory has been so successful that the people who are going to make progress are the people who will be building on this idea."

String Theory and Its Discontents

Stephen King's 1987 novel *The Tommyknockers* begins in the woods, where Roberta stumbles over a partly buried piece of metal. She attempts to pull it out, but it won't give, so she begins digging. Soon other people join her in a fanatical quest to unearth whatever lies hidden there: a large object of unknown purpose that extends deep into the ground. As they uncover more, the diggers' health deteriorates, but they develop a new set of skills, including telepathy and superior intelligence. The object turns out to be an alien spaceship. Some people die. The end.

King himself called *The Tommyknockers* "an awful book."[15] Maybe he has a point there, but it's a great metaphor for string theory: an alien object of unknown purpose deeply buried in mathematics, and an increasingly fanatical crowd of people with superior intelligence trying to get to the bottom of it.

They still don't know what string theory is. Even string theorists' best friend, Joseph Conlon, referred to it mysteriously as "a consistent structure of something."[16] And Daniele Amati, one of the field's founders, opined that "string theory is part of 21st-century physics that fell by chance into the 20th century."[17] I know, it was more convincing in the twentieth century.

But for all the controversy that surrounds string theory in the public sphere, within the physics community few doubt its use. Unlike vortex theory, the mathematics of string theory is deeply rooted in theories that demonstrably describe nature: quantum field theory and general relativity. So we are certain string theory has a connection to the real world. We also know string theory can be used to better understand quantum field theory. But whether string theory really is the sought-after theory of quantum gravity and a unification of the standard-model interactions, we still don't know.

Proponents like to point out that since string theory is *a* theory of quantum gravity and it connects to theories we know to be correct, it seems reasonable to hope that it is *the* theory of quantum gravity.

String theory is such a huge and beautiful body of mathematics that they can't believe nature would not choose this way.

For example, a widely used textbook on string theory, written by Lars Brink and Marc Henneaux, begins this way: "The almost irresistible beauty of string theory has seduced many theoretical physicists in recent years. Even hardened men have been swept away by what they can already see and by the promise of even more."[18] And John Schwarz, one of the founders of the field, recalls: "The mathematical structure of string theory was so beautiful and had so many miraculous properties that it had to be pointing toward something deep."[19]

On the other hand, mathematics is full of amazing and beautiful things, and most of them do not describe the world. I could belabor until the end of eternal inflation how unfortunate it is that we don't live in a complex manifold of dimension six because calculus in such spaces is considerably more beautiful than in the real space we have to deal with, but it wouldn't make any difference. Nature doesn't care. Besides, not only has nobody proved that string theory uniquely follows from general relativity and the standard model, but such a proof is not possible because—join me for the chorus—no proof is better than its assumptions.

And we can never prove the assumptions to be true. Consequently, string theory is not the only approach, neither to quantum gravity nor to unification; it's just that the other approaches take different assumptions as starting points, such as Garrett Lisi's E8 theory, which starts from the premise that nature should be geometrically natural. But besides string theory there are only a few approaches to quantum gravity that have grown to sizable research programs.

At the present time, the biggest competitor of string theory is loop quantum gravity. To prevent the problems that usually appear when one tries to quantize gravity, it identifies new dynamic variables to be made quantum, such as small loops in space-time (hence the name). The proponents of loop quantum gravity consider it more important to respect the principles of general relativity from the outset than to take into account the unification of standard-model forces. String theorists take the opposite point of view—they think that requiring a unification of all forces provides extra guidance.

Asymptotically safe gravity is probably the most conservative continuation of the theories we presently have. Researchers in this field argue that we're simply wrong to think there is a problem with quantizing gravity. If we look carefully, they claim, the usual quantization works just fine. In asymptotically safe gravity, the problems that appear with quantizing gravity are prevented because gravity gets weaker at high energies.

Causal dynamical triangulation tackles the problem by first approximating space-time with triangular shapes (hence the name) and then quantizing it. A lot of progress has been made with this idea in recent years, especially to describe the geometry of the early universe.

There are a few other approaches to quantum gravity, but these are the most popular ones right now.[20] Almost all attempts at quantizing gravity currently assume that the symmetries we have found in the standard model and general relativity already reveal part of the underlying structure. A different perspective entirely is that the symmetries we observe are themselves not fundamental but emergent.

Beauty Emerges

Xiao-Gang Wen is a professor of condensed matter physics at MIT. His discipline deals with systems composed of many particles that act together—think solids, liquids, superconductors, and all that. In the language of Chapter 3, condensed matter physics uses effective theories, valid only at low resolution or low energy—it's not part of the foundations of physics. But Xiao-Gang thinks that the universe and everything in it works like condensed matter.

In his imagination, space is made of tiny elementary units, and what we think of as particles of the standard model are merely collective motions—"quasi-particles"—of these elementary units. Xiao-Gang also has a quasi-particle that is a graviton, so gravity is covered too; he is after a full-blown theory of everything.[21]

In Xiao-Gang's universe, the elementary units are quantum bits or "qubits," the quantum variants of classical bits. A classical bit can take on two states (say, 0 and 1), but a qubit can be both at once, in any possible combination. In a quantum computer, a qubit is made of other particles. But in Xiao-Gang's theory, the qubits are fundamental. They aren't made of anything else, the same way string theory's strings aren't made of anything else—they just are. According to Xiao-Gang, the standard model and general relativity are not fundamental but emergent, and they emerge from qubits.

I think his idea will make a good antidote to the appeal of string theory, so I schedule a call with Xiao-Gang.

"What is it," I begin, "that you don't like about the existing approaches—string theory, loop quantum gravity, susy—when it comes to quantum gravity and unification?"

"I have this very strict view of a quantum theory," Xiao-Gang says, and begins to lay out his ideas. To describe the qubits and the way they interact, he uses a big matrix—a more formal version of a table—whose entries each describe a qubit. This matrix changes with an absolute time, thereby breaking the union of space and time that Einstein introduced.

I don't like that at all. But that was the point of talking to him, I remind myself.

"Your matrix is finite?" I ask, not quite willing to believe the whole universe can be written down in a table.

"Yes," he says, and adds that in his approach the universe itself is finite. "If we view the space to be a lattice, each lattice site will have one or a few qubits. We claim the lattice spacing might be at the Planck scale. But the lattice has no continuous geometry; the universe is just discrete qubits. The quantum dynamics of the qubits are described by a matrix, and the matrix is finite."

Wonderful, I think; it's even uglier than I expected. "You just postulate this matrix?" I ask him.

"Yes," he says, "and I believe all the major features of the standard model can be obtained from [it]. We have no complete model yet, but all the necessary ingredients can be produced by the qubits on the lattice."

"What happens to special relativity?"

"Exactly. This is something we should worry about indeed."

Indeed, I think, while he explains that special relativity "is consistent with the qubit approach but not naturally."

"What does 'not naturally' mean?" I want to know.

Xiao-Gang tells me that to get special relativity right, he must fine-tune the model's parameters. "Why it has to be that, I don't know," he says. "But if you insist, I can do that."

Once fine-tuned, Xiao-Gang's qubit model can approximately reproduce special relativity, or so he tells me.

"But the gauge symmetries are emergent at low energies?" I query, because I want to make sure.

"Yes," he affirms. While the "fundamental theory of nature may not have any symmetry," Xiao-Gang explains, "we don't need any symmetry in the qubit model to get gauge symmetry at low energies."

Furthermore, Xiao-Gang says that he and his collaborators have hints that the model might also contain an approximation to general relativity, though he emphasizes they don't yet have definite conclusions.

I am skeptical, but I tell myself to be more open-minded. Isn't this what I was looking for, something off the well-trodden path? Is it really any weirder to believe everything is made of qubits than of strings or loops or some 248-dimensional representation of a giant Lie algebra?

How patently absurd it must appear to someone who last had contact with physics in eleventh grade that people get paid for ideas like that. But then, I think, people also get paid for throwing balls through hoops.

"How has your work been received?" I ask.

"Not well," Xiao-Gang tells me. "The people from high energy [physics] don't care much what we try to do. They ask 'Why?' because they think the standard model plus perturbation theory is good enough and they say we don't need to go beyond that."

Suddenly I understand where Xiao-Gang is coming from. This isn't about unification at all. He wants to clean up the dirty math of the standard model.

ꝏꝏ

IN CASE I left you with the impression that we understand the theories we work with, I am sorry, we don't. We cannot actually solve the equations of the standard model, so what we do instead is solve them approximately by what is known as "perturbation theory."

For this, we first look at particles that don't interact at all to learn how they move when undisturbed. Next, we allow the particles to bump into each other, but only softly, so that they don't knock each other off their paths too much. Then we make successive refinements that take into account an increasing number of soft bumps, until the desired precision of the calculation is reached. It's like first drawing an outline and then adding more details.

However, this method works only when the interaction between the particles isn't too strong—the bumps aren't too violent—because otherwise the refinements don't get smaller (or aren't refinements, respectively). That's why, for example, it's hard to calculate how quarks combine to form atomic nuclei, because at low energies the strong interaction is strong indeed and the refinements don't get smaller. Luckily, since the strong interaction becomes weaker at higher energies, calculations for LHC collisions are comparably straightforward.

Even though the method works in some cases, we know the math will eventually fail because the refinements don't continue to get smaller forever. For the pragmatic physicist, a method that delivers correct predictions is just fine, regardless of whether mathematicians can agree on why it works. But as Xiao-Gang points out, fundamentally we don't understand the theory. This might be a case of missing mathematics, or it might hint at a deeper problem.[22]

Xiao-Gang Wen thinks that this problem doesn't get the attention it deserves, and that the lack of attention means a lack of experiments that could help sort things out. "We need new experiments that force people to confront the issue," he says, and suggests that we study the behavior of matter in the early universe in cases that can't be treated by the usual approximations.

He is right, I think, sorry that I misjudged his idea. I had entirely forgotten about these well-known problems. Nobody ever seems to mention them.

"When I try to write a paper for people in the field of quantum gravity and high-energy particle physics, even though I am originally from high-energy particle physics, I feel communication is very difficult," Xiao-Gang says. "The basic point of view and the starting point are very different. From our point of view, the starting point is just a lot of lowly qubits. But this leads to gauge symmetry, [fermions], et cetera. And the beauty is emergent. It's not popular. It's not mainstream."

IN BRIEF

- Problems of consistency are powerful guides toward new laws of nature. They are not questions of aesthetics.
- But even consistency problems can't be resolved purely by math and without experimental guidance, because formulating the problem itself depends on the assumptions that were accepted as true.
- Theoretical physicists are guilty of sweeping difficult questions under the rug and instead focusing on questions that are more likely to produce publishable results in a short period of time.
- The reason for the current lack of progress may be that we focus on the wrong questions.

9

The Universe, All There Is, and the Rest

In which I admire the many ways to explain why nobody sees the particles we invent.

Laws Like Sausages

If you think there have been more groundbreaking innovations recently than ever before, you're right. In a 2015 study, researchers from the Netherlands counted adjectives used in scientific papers and found that the frequency of the words "unprecedented," "groundbreaking," and "novel" increased by 2,500 percent or more from 1974 to 2014. Runners-up were "innovative," "amazing," and "promising," with an increase of more than 1,000 percent.[1] We're trying hard to sell our ideas.

Science is sometimes called the "marketplace of ideas," but it differs from a market economy most importantly in the customers we cater to. In science, experts only cater to other experts and we judge each other's products. The final call is based on our success at explaining observation. But absent observational tests, the most important property a theory must have is to find approval by our peers.

For us theoreticians, peer approval more often than not decides whether our theories will ever be put to a test. Leaving aside a lucky few showered with prize money, in modern academia the fate of an idea depends on anonymous reviewers picked from among our colleagues.[2] Without their approval, research funding is hard to come by. An unpopular theory whose development requires a greater commitment of time than a financially unsupported researcher can afford is likely to die quickly.

The other difference between the marketplace of consumer goods and the marketplace of ideas is that the value of a good is determined by the market, whereas the value of a scientific explanation is eventually determined by its use to describe observations—it's just that this value is often unknown when researchers must decide what to spend time on. Science, therefore, isn't a marketplace that creates its own value; rather, it's a forecasting platform to identify an external value. The function of the scientific community and its institutions is to select the most promising ideas and support them. This means, however, that in science, marketing prevents the system from working properly because it distorts relevant information. If everything is groundbreaking and novel, nothing is.

The need to assess which ideas are worth pursuing absent experimental support surfaced first in foundational physics because it's the research area where new hypotheses are most difficult to test. It's a problem, however, that sooner or later will arise also in other disciplines. As data become increasingly hard to come by, the lag between theory development and experimental test increases. And since theories are cheap and plentiful but experiments are expensive and few, somehow we have to select which theories are worth testing.

This doesn't mean our descriptions of nature are doomed to remain social constructs with no relation to reality. Science is a social enterprise, no doubt about it. But so long as we do experimental checks, our hypotheses are tied to observation. You could insist that experiments too are performed and evaluated by humans, and that is correct—science is "socially constructed" in the sense that scientists are people and work in collaboration. But if a theory works, it works, and calling it a social construct becomes a meaningless complaint.

It does mean, however, that we can get stuck on unchecked hypotheses and outdated ideas when starved for new data. We in foundational physics are the canary in the coal mine. And we better not sit listlessly on the ground, because social constructivists are watching closely, looking forward to the autopsy.

The canary isn't doing well. You'd think that scientists, with the professional task of being objective, would protect their creative freedom and rebel against the need to please colleagues in order to secure continued funding. They don't.

There are various reasons scientists play along. One is that those who can't stand the situation leave, and those who stay are the ones with few complaints—or maybe they just manage to convince themselves everything is all right.[3] Another reason is that thinking about how research works best takes time from doing research and is a competitive disadvantage. Several well-meaning friends have tried to dissuade me from writing this book.

But the main reason is that scientists trust in science. They're not worried. "It's the system," they say with a shrug, and then they tell themselves and everybody willing to listen that it doesn't matter, because they believe that science works, somehow, anyway. "Look," they say, "it's always worked." And then they preach the gospel of innovation by serendipity. It doesn't matter what we do, the gospel goes; you can't foresee breakthroughs, anyway. We're all smart people, so just let us do our thing and be amazed by the unpredictable spin-offs that will come about. Haven't you heard that Tim Berners-Lee invented the World Wide Web to help particle physicists share data?

"Anything goes" is a nice idea, but if you believe smart people work best when freely following their interests, then you should make sure they can freely follow their interests. And doing nothing isn't enough.

I've seen it in my own research area, and this book tells the story. But this isn't a problem merely in the foundations of physics. Almost all scientists today have an undisclosed conflict of interest between funding and honesty. Even tenured researchers are now expected to constantly publish well-cited papers and win grants, both of which require ongoing peer approval. The more peers approve, the better.

Being open about the shortcomings of one's research program, in contrast, means sabotaging one's chances of future funding. We're set up to produce more of the same.

You will do much better playing this game if you manage to convince yourself it's still good science. Evidently I have been unable to do this. "Laws, like sausages, cease to inspire respect in proportion as we know how they are made," John Godfrey Saxe quipped. He had in mind civil laws, but today the same can be said about the laws of nature.

Digging Darkness

In 1930, Wolfgang Pauli postulated the existence of a new particle—the neutrino—to account for unexplained missing energy in nuclear decay. He called it a "desperate remedy" and confided to his colleague, astronomer Walter Baade, "I have done a terrible thing today, something that no theoretical physicist should ever do. I have suggested something which can never be verified experimentally."[4] The neutrino was detected twenty-five years later.

Since Pauli's days, postulating particles has become the theoretician's favorite pastime. We have preons, sfermions, dyons, magnetic monopoles, simps, wimps, wimpzillas, axions, flaxions, erebons, cornucopions, giant magnons, maximons, macros, branons, skyrmions, cuscutons, planckons, and sterile neutrinos—just to mention the most popular ones. We even have unparticles. None of these have ever been seen, but their properties have been thoroughly studied in thousands of published research articles.

The first rule for inventing a new particle is that you need a good reason it hasn't yet been detected. For this, you can postulate either that too much energy is needed to produce it or that it interacts too rarely for the sensitivity of existing detectors, or both.

Storing away new particles at high energy is particularly fashionable in high-energy physics. It can take a lot of energy to produce a particle, either because the particle itself is very massive or because the particle is strongly bound and the bond must be broken to see the particle. The alternative option, explaining a lack of detection through a presumed weakness of interaction, is more popular in astrophysics because such particles make good candidates for dark matter. They are collectively referred to as the "hidden sector" of a theory.

ꝏꝏ

THE UNIVERSE hides something from us. We have known this since 1930 when Fritz Zwicky turned the one-hundred-inch Hooker telescope at the Coma cluster, a few hundred galaxies bound by their own gravitational pull. The galaxies move with an average velocity determined by the total mass which binds them together. Zwicky, to his surprise, found they moved much faster than their combined mass could explain. He suggested that the cluster contains additional unseen matter and called it *dunkle Materie*—dark matter.

It didn't remain the only oddity in the skies. When Vera Rubin forty years later studied the rotation of spiral galaxies, she noticed that the galaxies' outer stars orbited around the center faster than expected. It was the same in each of the more than sixty cases she looked at. The velocity a star needs to remain in a stable orbit depends on the total mass at the center of its motion, and that the galaxies' outer arms were rotating so fast meant the galaxies had to contain more matter than was visible. They had to contain dark matter.

As more experiments launched, evidence piled up that similar things were going on everywhere in the universe. The fluctuations of the cosmic microwave background fit the data only when we add dark matter. Dark matter is also needed to make the formation of galactic structures in the universe match our observations. Without dark matter, the universe just wouldn't look the way we observe it to be. Further evidence comes from gravitational lensing, the distortion of light caused by space-time curvature. Clusters of galaxies bend light more than their visible mass can account for. Something else must be there to bend space-time that strongly.

The first guess for this something else was that galaxies contain unexpectedly many hard-to-see stellar objects, like black holes or brown dwarfs. But these should also be present within our galaxy and cause frequent gravitational lensing, which hasn't been seen. The idea that dark matter consists of ultra-compact objects with masses much smaller than a typical star hasn't entirely been ruled out because these wouldn't bend light enough to cause observable gravitational lensing. It is, however, unclear how such objects would have formed to begin with. Physicists therefore presently favor a different type of dark matter.

The explanation that has attracted the most attention is that dark matter is made of particles which collect in clouds and hover around the visible disks of galactic matter in almost spherical halos. The known particles, however, almost all interact too strongly and clump too much to form such halos. The exceptions are neutrinos, but they are too light, move too fast, and don't clump enough. So, whatever particle dark matter is made of, it must be something new.

∞∞

THE SECOND rule for inventing a new particle is that you need an argument for why it's just about to be discovered, because otherwise nobody will care. This doesn't have to be a good argument—everyone in the business wants to believe you anyway—but you have to give your audience an explanation they can repeat. The common way to do this is to look for numerical coincidences and then claim they hint at new physics for a planned experiment, using phrases like "natural explanation" or "suggestive link." If your idea doesn't produce such a coincidence, don't worry—simply try the next idea. Just statistically you'll sometimes find a match.

A particularly lucky numerical coincidence that has driven much research in astrophysics is the "WIMP miracle." WIMP stands for "weakly interacting massive particle." These particles are presently the most popular candidates for dark matter, not least because they can easily be accommodated in supersymmetric theories. From their mass and interaction rate we can estimate how many WIMPs would have been produced in the early universe, and this gives about the correct abundance for dark matter, near the measured value of 23 percent. This is the relation known as the WIMP miracle.

According to astrophysicist Katherine Freese, the WIMP miracle is "a major reason why WIMPs are taken so seriously as dark matter candidates."[5] Jonathan Feng, also a well-known researcher in the field, finds this numerical match "particularly tantalizing," and others agree that it "has been the primary theoretical motivation in the field for many years."[6] It has also been a motivation for experiments.

Since we know the total mass that dark matter adds to the Milky Way, we can estimate how many dark matter particles of a certain individual mass must be drifting through us. The number is enormous: for a typical WIMP mass of 100 GeV, about 10 million WIMPs pass through the palm of your hand every second. But they interact very rarely: an optimistic estimate is that a kilogram of detector material interacts with one WIMP per year.[7]

But rarely isn't never. We might find evidence for the presence of WIMPs by closely monitoring large amounts of material that is shielded from all other particle interactions. Every once in a while, a dark matter particle would bump one of the material's atoms and leave behind a little bit of energy. Experimentalists currently search for three possible traces of this energy: ionization (knocking electrons off atoms, leaving behind a charge), scintillation (causing an atom to emit a flash of light), and phonons (heat or vibration). Such sensitive experiments are often located deep underground, where most particles from cosmic radiation are filtered out by the surrounding rock.

The possibility of looking for rare dark matter interactions was first pointed out by Mark Goodman and Edward Witten in 1985.[8] The search for dark matter particles began as an afterthought: experimentalists working with a detector originally developed to catch neutrinos reported in 1986 on the first "interesting bounds on galactic cold dark matter and on light bosons emitted from the sun."[9] In plain English, "interesting bounds" means they didn't find anything.

Various other neutrino experiments at the time also obtained interesting bounds.

In the early 1990s, dark matter got its first own experiment, COSME. Carried by the attention that supersymmetry attracted, numerous other detectors were deployed in rapid sequence: NaI32, BPRS, DEMOS, IGEX, DAMA, and CRESST-I.[10] Those derived further interesting bounds. In the mid-1990s, EDELWEISS obtained "the most stringent limit based on the observation of zero event."[11]

This, however, might merely mean that the sought-after particle interacts more weakly than anticipated. And so more experiments were commissioned: ELEGANTS, CDMS, Rosebud, HDMS, GEDEOn, GENIUS, GERDA, ANAIS, CUORE, XELPLin, XENON 10, and XMASS. CRESST I was upgraded to CRESST II. CDMS was upgraded to SuperCDMS. ZEPLIN I was upgraded to II and then to III. They all delivered interesting bounds. And new detectors with higher sensitivity went online: CoGeNT, ORPHEUS, SIMPLE, PICASSO, MAJORANA, CDEX, PandaX, and DRIFT. In 2013, XENON100—meanwhile probing an interaction probability smaller by a factor of 100,000 than the one originally considered—reported "no evidence for a dark matter signal."[12] XENON100 recently upgraded to XENON1T. Further experiments are under construction.

Thirty years have passed. Dark matter still hasn't been detected. The parameter range in which the WIMP miracle holds has meanwhile been excluded.[13]

Sitting Around Hoping

But science needs patience, I remind myself, and consult the Internet for youthful optimism and energy. Twitter offers me Katherine "Katie" Mack, better known as "Astrokatie."

Astrokatie recently rose to Twitter fame when one of her witty replies was retweeted by J. K. Rowling. I scroll down Katie's feed. There's a selfie with a cat, a photo with Brian Cox, a recording of an interview she gave on TV, and many, many replies to queries about astrophysics. When you need an informed opinion about water on Mars or the recent dark matter searches, you can count on Astrokatie. When you want facts about gravitational waves or about last week's exoplanet discovery, she's the one to ask. And when you look for someone to speak up on sexism and harassment in science, Astrokatie's there for you too: she uses her Twitter reach to call attention to the underrepresentation of minorities and women in science, something that many scientists prefer to forget.

Here's a modern scientist, I think, and send her a message.

Behind Katie's public persona is a professional astrophysicist and postdoctoral research fellow at the University of Melbourne, Australia. When we speak on Skype, Katie wears a jacket with the NASA logo, and nothing has ever looked cooler than that. I ask about her research area, and she sums it up with: "I have worked on a lot of things that dark matter probably isn't."

"What makes a model attractive?" I ask.

"For the stuff that I do, what makes a model interesting is does it have consequences that we can find," Katie says. "I approach theories in a very practical way. I am interested in really exciting new models, but only if they have consequences that can be tested.

"When I was doing my PhD I was working on axions. Axions have always been my favorite dark matter models because they play such an important role in so many things. They come out of [the strong nuclear force] and then they fit in with cosmology, and also with string theory. From a theory standpoint I think axions are the most attractive dark matter candidate because you don't have to make them out of nowhere; they come out from elsewhere, and that seems to be the best thing."

∞∞

After WIMPs, axions are the next most popular dark matter candidate. The original axion, invented to solve a fine-tuning problem with the strong nuclear force (the strong CP problem), was ruled out almost as quickly as it was proposed. The theory was then amended to make the axion interact only very weakly. Sometimes explicitly referred to as the "invisible axion," the new particle is hard to detect, but it also makes a good dark matter candidate.

Invisible axions, if they exist, would be produced in the early universe. They can be created in states of very low energy and then form a stable condensate that permeates the universe. This could be the dark matter that we observe. But axions don't have any miracle going for them, and if you want to get their density to match that of dark matter, then you have to fine-tune how they started in the early universe.[14] That makes them less appealing than WIMPs. Still, particle physicists like them because they prettify the strong nuclear force.

Many of the previously mentioned WIMP experiments are also sensitive to axions and have delivered interesting bounds. But unlike WIMPs, axions couple to electromagnetic fields, if only weakly, and can therefore be detected by setting up strong magnetic fields, which would turn a small fraction of the incoming axions into photons. By this method, the Axion Dark Matter Experiment (ADMX) has searched for dark matter axions since 1996. The CERN Axion Solar Telescope (CAST), which began taking data in 2003, looks out for axions produced in the Sun at energies much higher than the dark matter axions. Other experiments, like ALPS-I and ALPS-II, PVLAS, and OSQUAR, try to observe the reverse process—photons turning into axions—by closely studying the behavior of light in magnetic fields. So far, none of these experiments has found anything.

ꝏꝏ

KATIE SAYS: "Unfortunately, in my PhD work I couldn't find good ways to make axions work without some fine-tuning in one place or the other. That made it less fun. Because the problem the axion solves, the strong CP problem, is a fine-tuning problem. So if you make a new fine-tuning problem, you just push it into a new regime."

"What's wrong with fine-tuning?" I ask Katie.

"In general, fine-tuning tells us there is something unnatural about our theories. If there is a number that is really small, you have to assume that you got really, really lucky. And that's never appealing. It just seems that there should be some other explanation for it. In other cases, when there has been fine-tuning, there has always been some explanation and in the end it turned out it wasn't fine-tuning."

"You say you got really lucky, but how do you make a statement about probability without having a probability distribution?" I ask.

"Well, you can't. It's just a sense of aesthetics among theorists," Katie says. "If there is a small number, they don't like that. It's much more appealing to have a number close to 1. I don't know of any overarching reason that says it has to be that way, that we have all our constants of order 1. It's just the way that we approach our theories. It's not really Occam's razor because it doesn't make anything simpler, but it feels this way."

"It's something people do, but is it something we should be doing?"

Says Katie: "If you are basing the whole reason for existence of a new particle on solving a fine-tuning problem, then you actually have to solve it. But as someone who is not a model builder, I don't have to get so much into this. So I'm a little agnostic. But I do think that naturalness is just more appealing. And, building up on the theoretical physics that's been done before, it does look like guidance through simplicity is a good principle to start with. So I'm sympathetic to naturalness concerns. But I can't make big proclamations on it, and I don't have to. And that's good."

"Is the multiverse an option?" I ask.

"I hate the multiverse," Katie says. "I know it's a cliché to hate the multiverse, but I hate it. I suppose there's kind of an elegance to it, because you can just give up trying to understand the parameters. But I don't like that it's hard to test, and I find it ugly. It's dirty and messy and we just happen to be over here and you have to bring in anthropics...I really don't like it. I don't have a good argument against it, but I really don't like it."

"What do you make of the recent LHC data that seems to require a fine-tuning of the Higgs mass?"

"I'm not actively working on it, so I don't know what the challenges are and what the constraints are. I'm just sitting around hoping. But I'd be really surprised if it was just tons of epicycles on top of each other to get everything to work out. We have some indication that we need new physics, and supersymmetry doesn't look like it's going to work out. But I'm an optimist. I guess we'll find something to replace supersymmetry that isn't just horribly tuned.

"The Higgs mass," she says, "it worries me in the sense that I don't know what's going to happen. I've heard people say that this is the nightmare scenario. But I don't think we'll reach a dead end. I guess I'm just trusting the particle physicists to come up with new ideas and new models. I don't find it depressing; I find it exciting—now we have this big mystery.

"Maybe there is nothing more beautiful than the standard model and it gets more ugly after this," Katie says, "but my gut feeling is that we'll find a way to simplify the messier picture into something more unified. This is what I've always found appealing about physics: to reveal the beautiful picture that makes the messy things come together."

"Does it worry you that the dark matter detection experiments haven't found anything?" I ask.

"I'm not too worried that they haven't identified the particle," Katie replies. "I think the evidence that dark matter is a particle with certain properties has just gotten better. We have stronger constraints now and it gets harder to find and that's a bummer. But everything fits together with the picture that it's a particle. It's getting intriguing what this could be. I would be concerned about the possibility of dark matter not being a particle if there was a different model that fits the data as well."

"There are certainly people claiming modified gravity can do the same thing," I mention.

"It's not very convincing. I've never seen anything suggest that [particle dark matter] is a bad fit. And I've never seen anything suggest that modifying gravity is any better. I think that particles are simpler than adding [some] fields. If modified gravity would fit the data better, I'd be interested."

ꝏꝏ

TO EXPLAIN the existing cosmological data, we have to assume the universe contains two new, heretofore unexplained components. One of them is dark energy. The other component is usually attributed to particle dark matter, collectively described as a fluid. However, an effect just like this extra fluid could also come about because gravity's response to ordinary matter isn't what our equations predict.

But it is more difficult to modify gravity than it is to add particles to the standard model. A modification of gravity would be present everywhere, while particles could be over there in one amount and over here in another amount. Particle dark matter, therefore, is much more flexible.

Maybe it's too flexible. Astrophysicists have found regularities among galaxies that particle dark matter cannot account for, like the "Tully-Fisher relation"—an observed correlation between the brightness of a galaxy and the velocity of its outermost stars. And there are other problems with particle dark matter. It predicts, for example, that the Milky Way should have more satellite galaxies than have been observed, and it offers no explanation for why the ones we have observed almost lie in a plane. Also, the galactic centers aren't coming out correctly with particle dark matter; the matter density should be higher in the centers than we observe.

These shortcomings may be due to astrophysical processes that haven't yet been properly incorporated into the concordance model. That would be interesting, but it wouldn't change the foundations of physics. Or maybe these shortcomings tell us that particle dark matter isn't the right explanation.

The first attempt at modified gravity, known as modified Newtonian dynamics, did not respect the symmetries of general relativity, and this very much spoke against it.[15] Newer versions of modified gravity respect the symmetries of general relativity and still explain the observed galactic regularities better than particle dark matter.[16]

But they don't do well on distances far below or far above galactic size.

The solution might be somewhere in between. Recently a group of researchers has proposed that dark matter is a superfluid, which at short scales resembles modified gravity but at long scales allows the flexibility of particle dark matter.[17] This idea combines the benefits of both without the disadvantages of either.

And despite the different terminology, the mathematics of modified gravity and the mathematics of particle dark matter are almost the same. As Katie says, for modified gravity one adds new—so far unobserved—fields. For particle dark matter, one adds new—so far unobserved—particles. But particles are described by fields, so the difference is minuscule. What sets apart both approaches is the type of fields and the way that they interact with gravity. For a particle physicist, the fields of modified gravity have untypical properties. They are unfamiliar. They are not pretty. And so modified gravity has remained an idea on the fringe of the community. Modified gravity has merely several dozen supporters behind it, while thousands work on WIMPs and axions.

Presently, modified gravity cannot fit all the cosmological data as well as the concordance model. That might be because modifying gravity is just the wrong thing to do. Or maybe it's because fewer people are trying to make it fit.

∞∞

"WHAT DO you think?" Katie asks at the end of our conversation. "Do you think we'll find more beautiful, simpler models?"

It occurs to me then that no one else I've talked to asked for my opinion. And I am glad they didn't, because I wouldn't have had an answer.

But during my travels it has become clear to me that I am not missing a justification for why my colleagues rely on beauty. There just isn't any. As much as I want to believe that the laws of nature are beautiful, I don't think our sense of beauty is a good guide; in contrast, it has distracted us from other, more pressing questions. Like the one that Steven Weinberg pointed out: that we do not understand the emergence of the macroscopic world. Or, as Xiao-Gang Wen reminded me, that we do not understand quantum field theory. Or, as the issue of the multiverse and naturalness shows, that we do not understand what it means for a law of nature to be probable.

And so I tell Katie that, yes, I think nature has more beauty in store for us. But beauty, like happiness, can't be found by complaining about its absence.

Feeble Fields and Fifth Forces

There's yet another way to postulate new physics and then hide it, which is to introduce fields that either become relevant only at very long distances or in the very early universe, both of which are hard to test. Such inventions are acceptable today because they too explain numerological coincidences.

In general relativity, the cosmological constant (CC) is a free parameter. This means there is no deeper principle from which the constant can be calculated—it has to be fixed by measurement. The accelerated expansion of the universe shows that the CC is positive and that its value is related to an energy scale comparable to the mass of the heaviest known neutrino. That is, for particle physicists, it is a very small energy scale (see Figure 14).[18]

If the CC is nonzero, a space-time that does not contain any particles is no longer flat. The cosmological constant is therefore often interpreted as a vacuum with nonzero energy density and pressure.

General relativity doesn't tell us anything about the value of the CC. In quantum field theory, however, we can calculate the vacuum energy density—and it comes out to be infinitely large. But in the absence of gravity this doesn't matter: we never measure absolute energies anyway, we merely measure energy differences. In the standard model without gravity we can therefore use suitable mathematical procedures to remove the infinity and get a physically meaningful result.

In the presence of gravity, however, the infinite contribution becomes physically relevant because it would cause an infinite curvature of space-time. This clearly doesn't make sense. Further inspection luckily shows that the vacuum energy is unbounded only if one extrapolates the standard model up to infinitely high energies. And since we expect this extrapolation to break down at the Planck energy (at the latest), the vacuum energy should instead be a power of the Planck energy. That's better—at least it's finite. But still it's much too large to be compatible with observation. A cosmological constant that large would have ripped us apart or would have recollapsed the universe long ago.

However, we can simply choose the free constant in general relativity so that when it is added to the contribution from quantum field theory (whatever that is), the result agrees with observation. Hence, the expectation that the sum is somewhere at the Planck energy is—again—based on a naturalness argument. If we were able to do the calculation, so the story goes, we would be unlikely to find two large numbers that almost but not exactly cancel, leaving behind merely the small value we measure.

The cosmological constant is therefore not natural, to use physics-speak. It requires fine-tuning. Its small value is not beautiful. There's nothing wrong with this constant—it's just that physicists don't like it.

You'd think a constant would be the simplest assumption a theory can possibly have. But the belief that the value of the CC requires an explanation is an excuse for theoreticians to devise new laws of nature. Weinberg led the way for doing this with the anthropic principle, and part of the community is now busy inventing probability distributions for the multiverse. Another well-used way to explain the value of a constant is to make it dynamic, so that it can change over time. If set up nicely, the dynamic constant may prefer a small value, which supposedly explains something. Such generalized versions of the CC are referred to as dark energy.

If dark energy isn't just a CC, then the universe's acceleration changes slightly over time. There's no evidence for that. But there is an extensive literature on conjectured dark energy fields, like chameleon fields, dilaton fields, moduli, cosmons, phantom fields, and quintessence. Experiments are under commission.

And these are not the only invisible fields that cosmologists play with. There is also the inflaton field, the field used to puff up the early universe.

ꝏꝏ

INFLATION—THE UNIVERSE'S rapid expansion right after the big bang—is a courageous extrapolation into the past, back to the time when the density of matter was much higher than the densities we have probed.

To make predictions coming from inflation, however, one first has to specify what the inflaton—the field invented to make inflation happen—does. This requires giving the inflaton a potential energy, which will depend on several parameters. Once a potential is chosen, one can use inflation to calculate the distribution of the density fluctuations in the early universe. The result depends on the parameters in the potential, and for some of the simplest models the calculation fits well with observation.[19] The same inflation models are also in good agreement with other observed properties of the cosmic microwave background.[20]

Inflation, therefore, is useful to relate observed parameters to an underlying mathematical model. However, the predictions depend on the potential for the inflaton field. We could choose one potential that fits current data and revise as necessary, but that wouldn't keep cosmologists occupied. And so they industriously produce inflation models, for each of which they calculate predictions for measurements that haven't yet been made.

A census in 2014 counted 193 inflaton potentials, and that was only those with a single field.[21] But theoreticians' ability to mint models that predict any possible future observations has just shown they can't predict anything. These models are severely "underdetermined," as the philosophers have it; there isn't enough data to extrapolate from unambiguously.[22] One can construct a model from current measurements, but not reliably predict outcomes of future measurements.

The situation prompted Joe Silk to remark that "one can always find inflationary models to explain whatever phenomenon is represented by the flavour of the month."[23] And in a recent article for *Scientific American*, cosmologists Anna Ijjas, Avi Loeb, and Paul Steinhardt complained that due to the large number of models, "inflationary cosmology, as we currently understand it, cannot be evaluated using the scientific method."[24]

At the risk of scaring you, there's an infinite number of inflaton potentials left for further studies. And of course one can also use several fields or other fields. There's much room for creativity.

According to current theory assessment, this is high-quality research.

The Bedrock of the Sciences

It's mid-April and it's Wuppertal, some miles north of Cologne, Germany. I had expected a hotel or an institute building, but the address belongs to a family house in the outskirts. There's a little flower garden in front. Ivy climbs around the doorway. I ring the bell and a woman about my age opens the door.

"Yes?" she says.

"Hi," I say. "Um," I say. "I'm Sabine," I say.

She only looks confused.

"I'm looking for George Ellis?" I say.

She blinks at me once, twice.

"That's my father. But he's in Cape Town."

She asks me in and makes a phone call.

"Oh my God," George shouts through the phone from onboard a bus, "she's two weeks early!"

∞∞

IT'S THE end of April and it's Wuppertal. Same street, same house, same ivy. I ring the bell, hoping I didn't again spend four hours on the road just to startle a stranger and head back home.

To my great relief it's George who opens the door. "Hello," he says, then asks me in and shows me to a sunlit kitchen. Children's paintings decorate the walls.

George Ellis, professor emeritus at the University of Cape Town, is one of the leading figures in the field of cosmology. In the mid-1970s, together with Stephen Hawking, he wrote *The Large Scale Structure of Space-Time,* still a standard reference in the field.[25] Already in 1975 he studied the question of what can be tested in cosmology, long before the multiverse brought the issue to everyone's attention.[26] But George's interests aren't restricted to cosmology. He has also explored emergence in complex systems—not only in physics but also in chemistry and biology—and he isn't afraid of philosophy either. He likes to look at the big picture. But recently he doesn't like what he sees.

"What are you worried about?" I begin.

"There are physicists now saying we don't have to test their ideas because they are such good ideas," George says. He leans forward across the table and stares at me. "They're saying—explicitly or implicitly—that they want to weaken the requirement that theories have to be tested." He pauses and leans back, as if to make sure I understand the gravity of the situation. "To my mind that's a step backwards by a thousand years," he continues. "You have written about this. What you say is very similar to what I think: it's undermining the nature of science. I don't like this for reasons some of which are the same as yours and others that are quite different.

"The reasons that are the same, I think, is that science is having a difficult time out there, with all the talk about vaccination, climate change, GMO crops, nuclear energy, and all of that demonstrating skepticism about science. Theoretical physics is supposed to be the bedrock, the hardest rock, of the sciences, showing how it can be completely trusted. And if we start loosening the requirements over here, I think the implications are very serious for the others.

"But then there are some very different reasons why I'm interested in this, reasons that I think you're probably not so sympathetic to—that is, what are the limits of science in relation to human life?

What can science do and what can it not do? What can it say about human values, about worth and purpose? I think that's very important for the relation of science to the wider community."

How could I possibly not be sympathetic to this? I let him go on.

"[A lot] of the reasons people are rejecting science is that scientists like Stephen Hawking and Lawrence Krauss and others say that science proves God doesn't exist, and so on—which science cannot prove, but it results in a hostility against science, particularly in the United States.

"If you're in the Middle West USA and your whole life and your community is built around the church, and a scientist comes along and says 'Get rid of this,' then they better have a very solidly based argument for what they say. But David Hume already said 250 years ago that science cannot either prove or disprove the existence of God. He was a very careful philosopher, and nothing has changed since then in this regard. These scientists are sloppy philosophers.

"But this is probably not what you are interested in."

That scientists are sloppy philosophers isn't news to me—after all, I'm one myself. But, I say, "I didn't quite get what this has got to do with theory assessment."

"It's theory assessment in the sense that some scientists are making extended claims of what science is about—what it can prove and disprove," George explains. "If Lawrence Krauss comes along and says science disproves some aspect of religion, is this a scientific statement, or is it a philosophical statement? And if this is a scientific statement, what's the proof? They're claiming it's a scientific statement. So this is an area in which I think we're having differences.

"For example, Victor Stenger wrote a book a while ago saying that science disproves the existence of God," George says. "I was asked to review it and so I wrote: 'I opened this book with great anticipation, waiting to see what was the experimental apparatus that gave the result and what did the data points look like and was it a three-sigma or five-sigma result?' " He gives a short laugh. "Of course there is no such experiment. These are scientists who haven't understood basic philosophy, like the work of David Hume and Immanuel Kant.

"It's to do with theory assessment," George explains, "because science has nothing to say about philosophical issues but they are claiming it does. Science produces facts that are relevant for philosophical issues, but there's a boundary between science and philosophy which must be respected. And I've spent a lot of time thinking about it."

So have I, if for other reasons. Minding the boundary between science and philosophy, I think, could help physicists separate fact from belief. And I don't see a big difference between believing nature is beautiful and believing God is kind.

"To come back to physics," I say, "you are worried about where physics is going?"

"Yes... You must have looked at the book on multiverses by..." In search of a name he asks, "The string theorist from Columbia?"

"Brian Greene?"

"Yes. He has these nine multiverses. Nine! And he is using a slippery-slope argument. So on this side you have Martin Rees saying that the universe doesn't just end beyond our visual horizon, so in that sense it's a multiverse. Of course I agree. And a little further you have Andrei Linde's chaotic inflation with infinitely many bubble universes. And then still further over there you have the string theory landscape in which the physics in each of the bubbles is different. And even further you get Tegmark's mathematical multiverse. And then, far over there, you get people like [Nick] Bostrom saying that we live in a computer simulation. That's not even pseudoscience, it's fiction."

I say: "It's a modern version of the clockwork universe, basically. Then it was gears and bolts, now it's quantum computers."

"Yes," George says. "But you see, Brian Greene lists it as a possibility in his book. And when people write things like this as scientific possibilities, I wonder: to what level can you have faith in what they're thinking? It's just ridiculous! Then what else is it that you can't trust in what they say?"

I don't think we have remotely as many differences as George believes.

"This matters to me because there's a lot of trust involved in science," George continues. "Say the LHC, I have trust in the people who do the experiments. And the Planck collaboration, I have trust in them.* Trust matters a lot in science. And if there are people going about saying it's really possible that we live in a simulation, I can't trust them as scientists, or even as philosophers."

During experiments, the LHC creates about a billion proton-proton collisions per second. That's too much data to store even for CERN's computing capacity. Hence the events are filtered in real time and discarded unless an algorithm marks them as interesting. From a billion events, this "trigger mechanism" keeps only one hundred to two hundred selected ones.[27] We trust the experimentalists to do the right thing. We have to trust them, because not every scientist can scrutinize every detail of everybody else's work. It's not possible—we'd never get anything done. Without mutual trust, science cannot work.

That CERN has spent the last ten years deleting data that hold the key to new fundamental physics is what *I* would call the nightmare scenario.

"I'm not against the multiverse," George says. "I'm just against saying it's established science. If people say that, they want to loosen the requirement of testability.

"They've been following a good line of reasoning leading to the multiverse proposal. But they have so many steps now to get there from well-established physics. Each step seems a good idea, but they are all unchecked extrapolations from known physics. It used to be that you make a hypothesis and you check that step, then you make a further hypothesis and you check that step, and so on. Without that reality check we might go down the wrong path."

* Planck was a satellite mission by the European Space Agency in operation from 2009 to 2013 with the task of mapping the CMB temperature fluctuations. Some of Planck's data is still being analyzed by the collaboration.

"But that's because experimental checks are so difficult," I say. "And then what do we do to proceed?"

"I think we have to go back and start with some basic principles," George says. "One thing is that we need to rethink the foundations of quantum mechanics because under all of this is the measurement problem of quantum mechanics. When does a measurement really take place? It's when, say, a photon is emitted or absorbed. And how do you describe this? You use quantum field theory. But if you pick up any book on quantum field theory you find nothing about the measurement problem."

I nod. "They just calculate probabilities but never discuss how the probabilities turn into measurement outcomes."

"Yes. So we need to go back [and] rethink the measurement problem."

The other principle George recommends is to not work with infinities.

"If people speak about infinities, it raises all kinds of paradoxes," George says. "I actually wrote about this already in the 1970s.[28] The point I raised there was that DNA is a finite code, and so if the probability of life is nonzero, then in a large enough volume of space you will eventually have used every possible combination of genetic codes and eventually you get an infinite number of genetically identical twins. You see, if you have an infinite universe, as soon as a probability is not zero it gives you an infinite number of occurrences of everything possible happening.

"The issue of infinity is one of my touchstones," he continues. "Hilbert already wrote about the unphysical nature of infinity in 1925.[29] He said infinity is needed to complete mathematics but it doesn't appear anywhere in the physical universe. Physicists nowadays seem to think they can treat infinity as just another big number. But the central nature of infinity is quite unlike any finite number. It can never be realized no matter how long you wait or what you do—it is always beyond access."

He concludes: "So I think it should be a philosophical foundational principle that nothing physically real is infinite. There's no way

I can prove it—it may or may not be true. But we should use it as a principle."

I say, "What confuses me is that in other areas physicists do use the absence of infinity as a principle."

"Do use it?"

"Yes—when we have an infinity appearing in a function, we assume it's not physical," I explain. "But there's no good *mathematical* reason why a theory should not have infinities. It's a philosophical requirement turned into a mathematical assumption. People talk about it but never write it down. That's why I say it gets lost in math. We use a lot of assumptions that are based on philosophy, but we don't pay attention to them."

"Correct," George says. "The problem is that physicists have been put off philosophy by a certain branch of philosophers who spout nonsense—the famous Sokal affair and all of that. And there *are* philosophers who—from a scientific viewpoint—do talk nonsense. But nevertheless, when you are doing physics you always use philosophy as a background, and there are a lot of good philosophers—like Jeremy Butterfield and Tim Maudlin and David Albert—who are very sensible in terms of the relationship between science and philosophy. And one should form a good working relationship with them. Because they can help one see what are the foundations and what is the best way to frame questions."

The Philosophy of Gaps

In 1996, physics professor Alan Sokal submitted a spoof paper, titled "Transgressing the Boundaries: Toward a Transformative Hermeneutics of Quantum Gravity," to an academic journal, where it was accepted for publication. That the journal's reviewers and editors couldn't tell patent nonsense from academic writing traumatized philosophers and bolstered physicists' confidence in their own superiority.

The Sokal hoax is one of the reasons philosophers and physicists, especially those working on foundational questions, presently have a difficult relationship. I know many physicists who use the word "philosophy" as an insult, and even those with sympathy for the philosophers' quest doubt its use.

And understandably so. I've heard philosophers reinvent arguments physicists have long known to be wrong, I've heard philosophers worry about paradoxes physicists solved ages ago, and I've heard philosophers deduce how natural laws should be while ignoring how natural laws are. In short, there are unfortunately many philosophers who don't notice when they are out of their depth.

The same can be said of physicists, though, and I would be surprised if some philosophers did not go and say the same about me. Physicists draw on philosophical arguments more frequently than they like to admit, and I surely am no exception. It's easy enough for us to discard philosophy as useless—because it is useless.

Scientists are very goal-oriented in their professional pursuit, interested in acquiring new knowledge only if it has the chance to advance their research. But I yet have to find a philosopher who actually came up with something a physicist could have worked with. Even the philosophers who understand physics seem content with analyzing or criticizing what we do. Being useful isn't on their agenda.

In that, my experience has been, for all I can tell, typical. I can subscribe to Lawrence Krauss's summary: "As a practicing physicist... I, and most of the colleagues with whom I have discussed this matter, have found that philosophical speculations about physics and the nature of science are not particularly useful, and have had little or no impact upon progress in my field."[30] Steven Weinberg similarly remarked that "a knowledge of philosophy does not seem to be of use to physicists."[31] And Stephen Hawking has gone so far as to say that "philosophy is dead. Philosophy has not kept up with modern developments in science, particularly physics. Scientists have become the bearers of the torch of discovery in our quest for knowledge."[32]

One response to this has been that of Massimo Pigliucci, a philosopher at the City University of New York, who simply declared that "the business of philosophy [is] not to advance science."[33] Well, to the extent that the business of philosophy is not to advance science, it's not hard to see why scientists don't find it useful.

But clearly I shouldn't let one philosopher speak for a whole community. And so I was delighted to find that Tim Maudlin agrees that "physics needs philosophy" and that physicists stand to benefit because "philosophical skepticism focuses attention on the conceptual weak points in theories and in arguments."[34] Excellent. But, damn, where have you been? Where were you twenty years ago, ten years ago? Where have you been while we worked ourselves into this mess?

Today most problems in the foundations of physics are philosophical itches, not tensions with data, and we need philosophy to get to the bottom of our discomfort. Are numerological coincidences something we should pay attention to? It is ever justified to use aesthetic perception to assess laws of nature? Do we have any reason to believe that laws that are more fundamental should also be simpler? And if scientists churn out hypotheses by the hundreds to keep the presses going, what are good criteria to assess the promise of their ideas?

We need philosophers to bridge the gap between pre-scientific confusion and scientific argumentation. This also means, though, that when science progresses, when our knowledge expands, the room for philosophy inevitably shrinks. Like good psychologists, good philosophers of science succeed in making themselves superfluous. And like good psychologists, they shouldn't be offended if a patient furiously denies needing help.

ꝏꝏ

"I THINK THAT naturalness, that a theory should not have numerical coincidences, is also a philosophical criterion," I say.

"Yes."

"You say yes, but when I talk to people of my generation, many just treat it as a mathematical criterion. But if you want to make it into a mathematical criterion, you need a probability distribution. And where does that come from? Well, you need a theory for the probability distribution, or another probability distribution for the probability distribution of that theory, and so on."

"Yes," George says again. "And in the multiverse you can argue for any probability distribution you like, but cannot prove the probability distribution applies to the physics. [It's] not scientifically testable: it's an ad hoc theory to fit the data." He pauses for a moment, then adds: "I think the world of theoretical physics is in a very strange place."

IN BRIEF

- The current structure of academia strongly favors ideas that proliferate quickly and widely, such as beautiful but hard-to-test theories.
- Theoretical physicists now have well-established practices for producing new laws of nature that will remain untestable for long periods of time.
- Contact with philosophy may help physicists to identify which questions are worth asking, but there is presently little such contact.
- The reliance of theoretical physicists on criteria of beauty and the resulting lack of progress represent a failure of science to self-correct.

10
Knowledge Is Power

In which I conclude the world would be a better place if everyone listened to me.

I, Robot

It's hardly surprising that physicists find beauty in the laws of nature. If you sat through math class seeing equations as unsightly scribbles, you probably didn't go on to become a theoretical physicist. There aren't many physicists on record who complain that the laws of nature are appalling for the same reason there aren't many truck drivers who complain that big engines are ugly: we chose our profession because it appeals to us.

And of course seeking beauty remains a motivation throughout our work life: "Because it feels good to do the work and it keeps it exciting," as Gordy Kane said. Or, as Gian Francesco Giudice put it: "For me it is this unreasonable aspect that makes physics fun and exciting." And the prospect of finding beauty adds much to the excitement. As Dan Hooper writes in his book about supersymmetry: "The most beautiful theory could be written simply and briefly, perhaps even as a single equation.... As I am writing this paragraph, I can feel my heartbeat rise just a little and my palms begin to sweat."[1]

Yes, scientists are humans with sweaty palms, even though the public doesn't like to see us this way. In a recent survey scientists were judged more "trustworthy" than a "regular person," but also more "robot-like," "goal-oriented," and "cold."[2] Not very nice. But the insult isn't the judgment itself; the insult is mistaking the occupation for the person. As far as the occupation is concerned, the cliché is correct: scientists attempt the superhuman. We try to transcend the shortcomings of human cognition, and we do this by procedures meant to prevent us from lying to others and to ourselves.

The current procedures, however, are insufficient. To be good scientists we also have to be aware of our wishes, desires, and weaknesses. We have to be aware of our own humanity—and correct our shortcomings when necessary.

I'm sure you've heard the tale that we were all born as little scientists and discovered the world naturally until misguided education interfered with our instincts. I've heard it too. It's romantic, but it's wrong. Yes, we are born curious, and human infants learn quickly by trial and error. But our brains didn't develop to serve science; they developed to serve us. And what served us well during evolution does not always serve us well in science.

Math Piled on Top of Math

Physicists aren't the only scientists who chase after beauty. James Watson, for example, recalls that Rosalind Franklin was convinced that DNA was structured as a double helix because it was "too pretty not to be true."[3] Biologists preferentially study pretty animals.[4] And the mathematician David Orrell has argued that climate scientists favor elegant models to the detriment of accuracy.[5]

But chasing after beauty isn't a theory of everything. If there isn't a cognitive bias for trying to kill too many birds with one stone, there should be one, and I'll try not to fall for it. I've noticed aesthetic bias in only a few scientific disciplines, and even in physics it's dominant mainly in the areas I wrote about. Before I move on to the more general problem of social and cognitive biases, however, I want to highlight a case where the desire for elegant mathematics impacts our lives somewhat more than does quantum gravity: economics.

ꝏꝏ

THE ECONOMICS profession went astray because economists, as a group, mistook beauty, clad in impressive-looking mathematics, for truth." So argues the American economist Paul Krugman.[6] Like many theoretical physicists, I once considered switching to economics in the hope of better job opportunities. I wasn't impressed by the math, but I was stunned by the lack of data. I wouldn't have called economics beautiful, but it was arguably simple. Too simple, I thought. It's one thing to expect elementary particles to obey simple, universal laws. It's another thing entirely to expect the same of the human race.

Of course, I wasn't the first physicist to think that economics would benefit from more elaborate math. Doyne Farmer is one of the founders of econophysics, a discipline that applies mathematical methods developed in physics to problems in economics. With a background in chaos theory and nonlinear dynamics, Doyne is now director of the Complexity Economics Program at the Institute for New Economic Thinking at the Oxford Martin School. I call him to ask what he thinks of elegant economic theories.

"In economics, one of the strange things is that there is a template that theories follow," Doyne says. "The template is basically that you want a system in which agents selfishly maximize their preferences, and if you don't have that they'll say that the theory has no economic content. They have this concept that all theories have to come from that principle. And it's very strongly enforced, particularly in the so-called top journals.

"Economics has a very strong journal hierarchy," he explains. "There's about 350 journals, and they know the order—they know this one is number twenty, that one is number thirty, and so on. There's five top journals, and having a paper in one of them is a very big deal. One paper in one of these top journals can get you tenure at a good university."

"Wow," I say. "This is even worse than in physics."

"Yes, worse in the sense that it's a very high level of conformity. There are very strong criteria applied for how a paper should be written, regarding style, and manner of presentation, and the type of theory, whether it's connecting to the beliefs of the mainstream for what the theory should be. They view it as making elegant and beautiful arguments. I just don't happen to find these arguments that elegant and beautiful. I mean, they might be elegant in some sense, but I don't believe in the underlying framework. To me it seems a little silly, to be honest.

"And it's not like string theory," Doyne says. "In string theory, at least, you could argue that you get some nice, original mathematics out. But in economics it's not deep or original mathematics. It's like turning the crank on standard stuff of analysis. I don't think that economists are contributing to mathematics in an interesting way.

"It might turn out that they are right," he continues, "and at the end of the day they find what they are looking for. But in economics I think it's clear that mainstream models are at best partially successful. And I think that's because in economics in order to make progress they need to abandon some of these principles that are there because it allows them to get elegant results rather than to explain how the world is.

"Like equilibrium," Doyne says. "They like it because once you assume equilibrium it's simple to derive results. But on the other hand, if that's not how the world works, if that's not what's underlying economy, then the whole thing just ends up wasting time. And in many cases I think that's what it is." In economics, equilibrium refers to a steady state in which supply and demand are balanced and the value of goods is optimized.

"I recall reading some of your papers criticizing equilibrium theory, but that was a decade ago," I say.[7] "I was thinking—hoping, maybe—that it's been recognized since that it's too simplistic.

Economics must be terribly messy when it comes to data. It's not like physics, where I can at least understand they believe in simplicity. I'd expect that if you'd wanted to describe a real-world economy, then it must be a lot of computer-based analysis."

"I totally agree, and I'm trying to push into this direction, but it's a minority."

"This surprises me," I say, "because 'big data' is a such a big word now, and I was thinking that economics should be the first to go for it."

"Economists have their ears pricked up," Doyne says. "But it's quite different from what they were doing in the past and it will take some time."

"How come economists got so hooked on mathematics?" I ask.

"Mathematics is in physics for a good reason," he explains. "It allows you to take a set of assumptions and reason to the consequences from these assumptions. So the idea to use mathematics in the social sciences is very exciting. The problem isn't with mathematics per se. The things that have become dominant have, I worry, become dominant because they are elegant and because people can prove conclusions, not because they are any good for describing the world. It's the mathematician looking under the lamppost for their keys.

"I think beauty is great, and I'm all for elegance," Doyne says. "But I worry when there is too much math piled on top of math piled on top of math, even if it's elegant math. It might be nice math, but at the end of the day how certain are you that you're really doing science?"

Don't Trust Me, I'm a Scientist

Have you heard the curious story of the neutron's life? A neutron is built of three quarks, and together with protons, neutrons make up atomic nuclei. Atomic nuclei are luckily stable, but take the neutron out of the nucleus and that neutron will decay, with an average lifetime of about 10 minutes. More precisely, 885 seconds plus or minus 10. The curious part is the plus or minus.

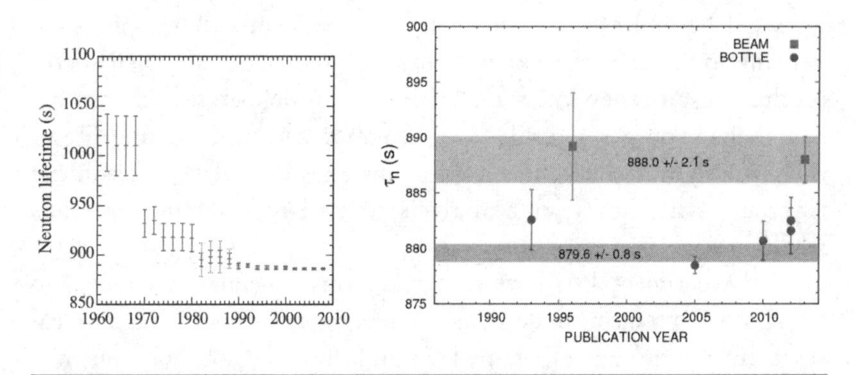

FIGURE 15. Neutron lifetime measurements by year. *Sources:* Patrignani C et al. (Particle Data Group). 2016. "Review of particle physics." *Chin Phys C* 40:100001. (Error bars are 1σ.) Bowman JD et al. 2014. "Determination of the free neutron lifetime." arXiv:1410.5311 [nucl-ex].

The neutron's lifetime has been measured with steadily increasing accuracy starting in the 1950s (Figure 15, left). Presently there are two different measurement techniques, which yield different results (Figure 15, right). This discrepancy is larger than what the measurement uncertainties allow for, which means the odds are less than 1 in 10,000 that it's due to chance.[8] This is a puzzling situation and might be a harbinger of new physics. But this isn't what I wanted to draw your attention to.

Look again at the left graph in Figure 15. The small middle ticks are the measurement values, and the vertical lines are the reported uncertainty. Note how the data come in steps and then suddenly make a jump to a new value that was previously strongly disfavored, sometimes even outside the reported uncertainty. The experimentalists seem to have not only underestimated the measurement uncertainties but also preferably arrived at best-fit values that reproduced previous results. And this isn't the only quantity that has seen such measurement jumps over time. Similar jumps have occurred in the last decades for at least a dozen lifetimes, masses, and scattering rates of other particles.[9]

We'll never know for sure why this happened. But a plausible explanation is that the experimentalists unconsciously tried to reproduce results they knew of. I don't mean deliberate fudging. It's just that if you get a result that disagrees with the existing literature, you're more likely to look for mistakes than if your result fits right in. This skews your analysis toward reproducing previous results.

But experimentalists have recognized this issue and taken steps to address such problems. Indeed, as you see, the results of the recent measurements do *not* agree (Figure 15, right). In many collaborations it is now common to settle on a method of analysis before even looking at the data (before the data is "unblinded") and then just follow through with the previously agreed-upon procedure. This can prevent the tendency to try out different analysis methods if the result isn't as desired.

In the life sciences too, the recent reproducibility crisis has increased efforts to guard against bias in experimental design, statistical analysis, and publication practices.[10] There's still a long way to go, but at least it's a start.

Here in theory development, the apparatus we work with is our brain. But we're not doing anything to avoid bias in its operation. We can't visualize our progress in simple graphs, but I'm sure that if we could, we'd also see among theorists a preference for reproducing existing results. On some topics we've piled up so many papers they have become self-supportive research areas even in the absence of experimental evidence. These are elaborate theoretical constructs that are being thoroughly tested—tested for mathematical consistency. Putting forward a different solution that is also mathematically consistent would amount to a result that disagrees with the existing literature.

Take black hole evaporation. There are no data. The firewall paradox (Chapter 8) showed that the most-studied attempt to solve the problem of black hole information loss—the gauge-gravity duality—violates the equivalence principle. It therefore doesn't solve the problem it was meant to solve because it's incompatible with the main postulate of general relativity.[11] But there is so much prior work in favor of this supposed solution, it's unthinkable to discard it. Instead, theoretical physicists are now trying to make the new result compatible with earlier work by reinventing quantum mechanics.

For example, Juan Maldacena and Leonard Susskind have postulated that entangled particles are linked by wormholes, deformations of space-time so strong that two formerly distant places become connected by a short tunnel.[12] Nonlocality then is no longer "spooky," but a property of space and time and our universe is pierced by wormholes. Wormholes would in particular connect the particle pairs of the Hawking radiation, removing both the firewall and the black hole information loss problem. This idea is developed in a space-time with a negative cosmological constant, so it doesn't describe the universe we live in. But they hope it is a general principle that also holds in our universe.[13]

They may be right. It's an exciting new idea. If nature works like that, it would be an amazing insight. And it fits so nicely with all the previous results.

This Tent Stinks

Cherishing beauty and wanting to fit in are human traits. But they distort our objectivity. These are cognitive biases that prevent science from working optimally, and they're biases that are presently unaccounted for. These are not the only biases you find among theoreticians. Where experimentalists go to great lengths to account for statistical biases, theoreticians proceed entirely undisturbed, happily believing it is possible to intuit the correct laws of nature.

Human cognitive biases aren't generally a bad thing. Most of them developed because they are, or at least have been, of advantage to us. We are, for example, more likely to put forward opinions that we believe will be well received by others. This "social desirability bias" is a side effect of our need to fit into a group for survival. You don't tell the tribal chief the tent stinks if behind you stand a dozen fellows with spears. How smart of you. But while opportunism might benefit our survival, it rarely benefits knowledge discovery.

Probably the most prevalent brain bug in science is confirmation bias. If you search the literature for support of your argument, there it is. If you look for a mistake because your result didn't match your expectations, there it is. If you avoid the person asking nagging questions, there it is. Confirmation bias is also the reason we almost always end up preaching to the choir when we lay out the benefits of basic research. You knew that without discovering fundamentally new laws of nature, innovation would eventually run dry, didn't you?

But there are other cognitive and social biases affecting science that are not as well-known.[14] Motivated cognition is one of them.[15] It makes us believe positive outcomes are more likely than they really are. Do you recall hearing that the LHC is likely to see evidence for physics beyond the standard model? That these experiments likely will find evidence for dark matter in the next couple of years? Oh, they are still saying that?

Then there is the sunk cost fallacy, more commonly known as throwing good money after bad. The more time and effort you've spent on supersymmetry, the less likely you are to call it quits, even though the odds look worse and worse. We keep on doing what we've been doing long after it's stopped being promising, because we already invested in it, and we don't like to admit we might have been wrong. It's why Planck quipped, "Science advances one funeral at a time."[16]

The in-group bias makes us think researchers in our own field are more intelligent than others. The shared information bias is why we keep discussing what everyone knows but fail to pay attention to information held only by a few people. We like to discover patterns in noise (apophenia). We think arguments are stronger if the conclusion seems plausible (belief bias). And the halo effect is the reason you are more interested in what a Nobel Prize winner says than what I say—regardless of the topic.

There's also the false consensus effect: we tend to overestimate how many other people agree with us and how much they do so. And one of the most problematic distortions in science is that we consider a fact to be more likely the more often we have heard of it; this is called attentional bias or the mere exposure effect. We pay more attention to information especially when it is repeated by others in our community. This communal reinforcement can turn scientific communities into echo chambers in which researchers repeat their arguments back to each other over and over again, constantly reassuring themselves they are doing the right thing.

Then there is the mother of all biases, the bias blind spot—the insistence that we certainly are not biased. It's the reason my colleagues only laugh when I tell them biases are a problem, and why they dismiss my "social arguments," believing they are not relevant to scientific discourse. But the existence of these biases has been confirmed in countless studies. And there is no indication whatsoever that intelligence protects against them; research studies have found no links between cognitive ability and thinking biases.[17]

Of course, it's not only theoretical physicists who have cognitive biases. You can see these problems in all areas of science. We're not able to abandon research directions that turn out to be fruitless; we're bad at integrating new information; we don't criticize our colleagues' ideas because we are afraid of becoming "socially undesirable." We disregard ideas that are out of the mainstream because these come from people "not like us." We play along in a system that infringes on our intellectual independence because everybody does it. And we insist that our behavior is good scientific conduct, based purely on unbiased judgment, because we cannot possibly be influenced by social and psychological effects, no matter how well established.

We've always had cognitive and social biases, of course. They are the reason scientists today use institutionalized methods to enhance objectivity, including peer review, measures for statistical significance, and guidelines for good scientific conduct. And science has progressed just fine, so why should we start paying attention now? (By the way, that's called the status quo bias.)

Larger groups are less effective at sharing relevant information.[18] Moreover, the more specialized a group is, the more likely its members are to hear only what supports their point of view. This is why understanding knowledge transfer in scientific networks is so much more important today than it was a century ago, or even two decades ago. And objective argumentation becomes more relevant the more we rely on logical reasoning detached from experimental guidance. This is a problem that affects some areas of theoretical physics more than any other field of science; hence the focus of this book.

Experimentalists push their own agenda, of course. They fancy the development of new technologies and don't leave decisions about future experiments up to theorists. But still, it's up to us theorists to point at new regions of parameter space worth exploring. We have a responsibility to assess our theories as objectively as possible in order to help identify the most promising new experiments.

Regardless of the field, as long as theories are developed by humans, the default assumption must be that theory assessment is both cognitively and socially biased unless steps are taken to address these issues. But no such steps are currently being taken. Hence, progress is almost certainly slower than it could be.

How could we have ended up in a situation like that? Because it's easy for us scientists to blame governing bodies or funding agencies, and there is no shortage of complaints: *Nature* and *Times Higher Education* seem to publish a rant about nonsensical attempts to measure scientific success every other week. When I share these articles on Facebook, they are guaranteed to get the thumbs-up. And yet nothing ever changes.

Complaining about others hasn't helped because it's a problem we've caused ourselves—and one that we must solve ourselves. We have failed to protect our ability to make unbiased judgments. We let ourselves be pushed into a corner, and now we are routinely forced to lie if we want to continue our work. That we accept this situation is a failure of the scientific community, and it's our responsibility to get this right.

It's not very popular to criticize your own tribe. But this tent stinks.

∞∞

CRITICISM IS cheap, say the critics. I spent nine chapters making a case that theoretical physicists are stuck on beauty ideals from the past, but now you may be wondering what else I think they should do. Don't I have an alternative to offer?

I don't have a miracle cure for the problems theoretical physicists are trying to solve, and if I told you I did, you'd be well-advised to laugh me off. These are tough problems, and complaining about aesthetic biases won't just make them go away. In the next section I offer some thoughts on where to start. But of course I have my personal preferences like everybody else. And of course I too am biased.

My intention here is more general and goes beyond my own discipline. Cognitive and social biases are a threat to the scientific method. They stand in the way of progress. While we will never be able to get rid of human biases entirely, it's not a situation we just have to accept. At least we can learn to recognize problems and avoid reinforcing them by bad community practices. In Appendix C I have collected some practical suggestions.

Lost in Math

Math keeps us honest, I told you. It prevents us from lying to ourselves and to each other. You can be wrong with math, but you can't lie. And it's true—you can't lie with math. But it greatly aids obfuscation.

Do you recall the temple of science, in which the foundations of physics are the bottommost level, and we try to break through to deeper understanding? As I've come to the end of my travels, I worry that the cracks we're seeing in the floor aren't really cracks at all but merely intricate patterns. We're digging in the wrong places.

As you have seen, most of the problems we presently study in the foundations of physics are numerological coincidences. The fine-tuning of the Higgs mass, the strong CP problem, the smallness of the cosmological constant—these are not inconsistencies; they are aesthetic misgivings.

But in the history of our field, mathematical deduction led the way only if we indeed had a problem of inconsistency. The inconsistency of special relativity with Newtonian gravity gave rise to general relativity.

The inconsistency between special relativity and quantum mechanics led to quantum field theory. The breakdown of the probabilistic interpretation of the standard model allowed us to conclude that the LHC must find new physics, which appeared in form of the Higgs boson. These were questions that could be tackled with math. But most of the problems we deal with now are not of this kind. The one exception is the quantization of gravity.

The first lesson I draw, therefore, is this: If you want to solve a problem with math, first make sure it really is a problem.

Theoretical physicists pride themselves on their experience and intuition. And I am all in favor of using intuition by making assumptions that may only later become justified (or not). But we have to keep track of these assumptions, or else we risk them becoming accepted even though they are unjustified. Intuition-based assumptions are often pre-scientific and fall into the realm of philosophy. If so, we need contact with philosophers in order to understand how our intuitions can be made more scientific.

Because of this my second lesson is: State your assumptions.

Naturalness is such an assumption. So is simplicity; reductionism does not imply a steady increase of simplicity toward smaller scales. Instead, we might have to go through a phase (in the sense of scales) where our theories become more complicated again. The reliance on simplicity, dressed up as unification or the decrease of the number of axioms, might mislead us.

But even with good problems and clearly stated assumptions, there still can be many mathematically possible solutions. In the end the only way to find out which theory is correct is to check whether it describes nature; non-empirical theory assessment will not do. In the search for a theory of quantum gravity and for a better formalism of quantum physics, the only way forward is to derive and test different predictions.

And so, my third and final lesson is this: Observational guidance is necessary.

Physics isn't math. It's choosing the right math.

The Search Goes On

June 22, 2016: The first rumors appear that the diphoton bump is fading away with the new LHC data.

July 21, 2016: The LUX dark matter experiment concludes its search and reports no signal of WIMPs.

July 29, 2016: The rumor that the diphoton anomaly is gone heats up.

August 4, 2016: The new LHC data are published. They confirm that the diphoton bump is gone for good. In the eight months since its "discovery," more than five hundred papers were written about a statistical fluctuation. Many of them were published in the field's top journals. The most popular ones have already been cited more than three hundred times. If we learn anything from this, it's that current practice allows theoretical physicists to quickly invent hundreds of explanations for whatever data happen to be thrown at them.

In the weeks that follow, Frank Wilczek loses his bet with Garrett Lisi that supersymmetry would be found at the LHC. A similar bet made at a conference in 2000 is settled in favor of the no-susy party.[19]

Meanwhile, I win a bet with myself by banking on my colleagues' continued failure while I finish writing. The odds were in my favor—they spent thirty years trying the same thing over and over again, expecting different results.

In October, the CDEX-1 collaboration reports they haven't seen any axions.

How long is too long to wait for a theory to be backed up by evidence? I don't know. I don't think this question even makes sense. Maybe the particles we are looking for are just around the corner and it's really only a matter of technological sophistication to find them.

But whether or not we will find something, it is already clear that the old rules for theory development have run their course. Five hundred theories to explain a signal that wasn't and 193 models for the early universe are more than enough evidence that current quality standards are no longer useful to assess our theories. To select promising future experiments, we need new rules.

In October 2016 the KATRIN experiment in Karlsruhe, Germany, begins operations. Its task is to measure the heretofore unknown absolute masses of neutrinos. In 2018, the Square Kilometer Array, a radio telescope under construction in Australia and South Africa, will begin searching for signals from the earliest galaxies. In the coming years, the g-2 experiment at Fermilab in Chicago and the J-PARC experiment in Tokyo will measure the magnetic moment of the muon to unprecedented precision, probing a long-standing tension between experiment and theory. The European Space Agency has tested grounds for the space-borne laser interferometer eLISA that could measure gravitational waves in unexplored frequency ranges, delivering new details of what happens during inflation. Much of the LHC data is yet to be analyzed, and we still might find signs of physics beyond the standard model.

We know that the laws of nature we presently have are incomplete. To complete them, we have to understand the quantum behavior of space and time, overhauling either gravity or quantum physics, or maybe both. And the answer will without doubt raise new questions.

Physics, it might seem, was the success story of the last century, but now is the century of neuroscience or bioengineering or artificial intelligence (depending on whom you ask). I think this is wrong. I got a new research grant. There's much work to do. The next breakthrough in physics will occur in this century.

It will be beautiful.

ACKNOWLEDGMENTS

I am grateful to all those whose interviews allowed me to create such a lively picture of the community: Nima Arkani-Hamed, Doyne Farmer, Gian Francesco Giudice, Gerard 't Hooft, Gordon Kane, Michael Krämer, Garrett Lisi, Katherine Mack, Keith Olive, Chad Orzel, Joe Polchinski, Steven Weinberg, Xiao-Gang Wen, and Frank Wilczek. Thank you so much—you were awesome!

I owe much to the many people who over the years helped me to better understand various topics this book touches on: Howie Baer, Xavier Calmet, Richard Dawid, Richard Easther, Will Kinney, Stacy McGaugh, John Moffat, Florin Moldoveanu, Ethan Siegel, David Spergel, Tim Tait, Tilman Plehn, Giorgio Torrieri, and countless others from whose seminars, lectures, books, and papers I have benefitted.

I also thank the volunteers who read drafts of this manuscript: Niayesh Afshordi, George Musser, Stefan Scherer, Lee Smolin, and Renate Weineck.

Special thanks go to Lee Smolin for finally realizing that he wouldn't be able to talk me out of writing this book, and to my agent, Max Brockman, and the people at Basic Books, especially Leah Stecher and Thomas Kelleher, for their support.

Finally, I want to thank Stefan for tolerating two years of curses about "the damned book," and Lara and Gloria for the much-needed distraction.

This book is dedicated to my mom who, when I was ten, let me wreck her typewriter. Look, Ma, it was finally good for something.

ACKNOWLEDGMENTS

APPENDIX A: THE STANDARD MODEL PARTICLES

The particles of the standard model (see Figure 6) are classified under the gauge symmetries.[1] The fermions of the strong nuclear force are the quarks, of which we have found six. They are called the up, down, strange, charm, bottom, and top quarks. The up, charm, and top quarks have a fractional electric charge of 2/3; the other three quarks have an electric charge of –1/3. The quarks' interaction is mediated by eight massless gluons, which are the gauge bosons of the strong force. Their number follows from the symmetry group of the strong force, which is SU(3).

The remaining fermions do not participate in the strong interaction and are called leptons. Of these we have also six: the electron, muon, and tau (each with electric charge –1) and their associated neutrinos, the electron neutrino, muon neutrino, and tau neutrino (which are electrically neutral). The electroweak interaction is mediated by the massless, neutral photon and by the massive Z, W^+, and W^- bosons, which have electric charge 0, +1, and –1, respectively. Again, the number of gauge bosons follows from the symmetry group, which for the electroweak interaction is SU(2) × U(1).

The fermions are subdivided into three generations that, roughly speaking, order them by mass. More important, though, the generations pack the fermions into sets that must contain equal numbers of quarks and leptons, otherwise the standard model wouldn't be consistent. The number of generations is not fixed by consistency requirements, but existing data strongly indicate there are only three.[2]

Besides the fermions (quarks plus leptons) and gauge bosons, there is only one more particle in the standard model, which is the Higgs boson. It is massive and is not a gauge boson. The Higgs is electrically neutral and its task is to give mass to the fermions and the massive gauge bosons.

Disappointed it's so ugly?

APPENDIX B: THE TROUBLE WITH NATURALNESS

The assumption of a uniform distribution is based on the impression that it's intuitively a simple choice. But there is no mathematical criterion that singles out this probability distribution. Indeed, any attempt to do so merely leads one back to the assumption that some probability distribution was preferable to begin with. The only way to break this circle is to just make a choice. Hence naturalness is fundamentally also an aesthetic criterion.[1]

A first attempt to justify the uniform probability distribution for the naturalness criterion might be to say that it doesn't introduce additional parameters. But of course it does: it introduces the number 1 as a typical width. "Oh," you say, "but the number 1 is the only number that naturalness allows me to use." Well, that depends on how you define naturalness. And you have defined naturalness by comparing the occurrence of a number to random chance. And what's the random distribution for that? And so it goes in a circle.

To better see why this criterion is circular, think of a probability distribution on the interval from 0 to 1 that is peaked around some value with a width of, say, 10^{-10}. "There," you exclaim, "you have introduced a small number! That's fine-tuned!" Not so fast. It's fine-tuned according to a uniform probability distribution. But I'm not using a uniform distribution; I'm using a sharply peaked one. And if you use this distribution, then it is very likely that two randomly selected numbers are at a distance of 10^{-10}. "But," you say, "that's a circular argument." Right, but that was my point, not yours. The sharply peaked probability distribution justifies itself as much or as little as the uniform distribution does. So which one is better?

Yes, I know, it somehow feels like a constant function is special, like it's somehow simpler. And it somehow feels like 1 is a special number. But is this a mathematical criterion or is it an aesthetic one?

You can try to take a meta approach to this problem and ask yourself whether there is a most likely probability distribution. For this you'll need a probability distribution in the space of probability distributions, and so on, leading to a recursion relation. The number 1 is indeed special in that it's the unit element of the multiplication group. One can hence try to construct a recurrence relation that converges to a distribution with a width of 1 as a limiting case. I've been toying with this idea, and to make a long story short, the answer is no: you always need additional assumptions to select a probability distribution.

For the experts, a slightly longer answer that illustrates the problem: There are infinitely many bases in the space of functions, none of which is preferred on mathematical grounds. We just happen to be very used to monomials, hence our liking for constant, linear, or quadratic functions. But you could equally well go for a uniform distribution in Fourier-space (of whatever parameter you are dealing with). Indeed, there is a uniform distribution for any possible basis that you can choose and they are all different. So if you want to use a recursion, you can swap the choice of distribution for a choice of the basis of distribution functions, but in either case you still have to make a choice. (The recursion would also bring in additional assumptions.)

Either way, whether or not you trust my conclusion that naturalness is not a mathematical criterion but an aesthetic one that's gotten "lost in math," it should give you pause that the question of how to pick the probability distribution is not discussed in the literature—even though one of the first papers to introduce the measures carefully pointed out that "any measure of fine tuning that quantifies naturalness" requires a choice that "necessarily introduces an element of arbitrariness to the construction."[2]

A modern incarnation of technical naturalness makes use of Bayesian inference. In this case the choice is moved from the probability distribution to the priors.[3]

APPENDIX C: WHAT YOU CAN DO TO HELP

Both bottom-up and top-down measures are necessary to improve the current situation. This is an interdisciplinary problem whose solution requires input from the sociology of science, philosophy, psychology, and—most importantly—the practicing scientists themselves. Details differ by research area. One size does not fit all. Here is what you can do to help.

As a scientist

LEARN ABOUT SOCIAL AND COGNITIVE BIASES. Become aware of what they are and under which circumstances they are likely to occur. Tell your colleagues.

PREVENT SOCIAL AND COGNITIVE BIASES. If you organize conferences, encourage speakers to list not only motivations but also shortcomings. Don't forget to discuss "known problems." Invite researchers from competing programs. If you review papers, make sure open questions are adequately mentioned and discussed. Flag marketing as scientifically inadequate. Don't discount research just because it's not presented excitingly enough or because few people work on it.

BEWARE THE INFLUENCE OF MEDIA AND SOCIAL NETWORKS. What you read and what your friends talk about affects your interests. Be careful what you let into your head. If you consider a topic for future research, factor in that you might have been influenced by how often you have heard others speak about it positively.

BUILD A CULTURE OF CRITICISM. Ignoring bad ideas doesn't make them go away; they will still eat up funding. Read other researchers' work and make your criticism publicly available. Don't chide colleagues for criticizing others or think of them as unproductive or aggressive. Killing ideas is a necessary part of science. Think of it as community service.

SAY NO. If a policy affects your objectivity, for example because it makes continued funding dependent on the popularity of your research results, point out that it interferes with good scientific conduct and should be amended. If your university praises its productivity by paper counts and you feel that this promotes quantity over quality, say that you disapprove of such statements.

As a higher ed administrator, science policy maker, journal editor, or representative of a funding body

DO YOUR OWN THING. Don't export decisions to others. Don't judge scientists by how many grants they won or how popular their research is—these are judgments by others who themselves relied on still others. Make up your own mind. Carry responsibility. If you must use measures, create your own. Better still, ask scientists to come up with their own measures.

USE CLEAR GUIDELINES. If you have to rely on external reviewers, formulate recommendations for how to counteract biases to the extent possible. Reviewers should not base their judgment on the popularity of a research area or a person. If a reviewer's continued funding depends on the well-being of a certain research area, that person has a conflict of interest and should not review papers in that area. That will be a problem because this conflict of interest is presently everywhere. See next three points to alleviate it.

MAKE COMMITMENTS. You have to get over the idea that all science can be done by postdocs on two-year fellowships. Tenure was institutionalized for a reason, and that reason is still valid. If that means fewer people, then so be it. You can either produce loads of papers that nobody will care about ten years from now, or you can be the seed of ideas that will still be talked about in a thousand years. Take your pick. Short-term funding means short-term thinking.

ENCOURAGE A CHANGE OF FIELD. Scientists have a natural tendency to stick to what they already know. If the promise of a research area declines, they need a way to get out; otherwise you'll end up investing money in dying fields. Therefore, offer reeducation support, one- to two-year grants that allow scientists to learn the basics of a new field and to establish contacts. During that period they should not be expected to produce papers or give conference talks.

HIRE FULL-TIME REVIEWERS. Create safe positions for scientists who specialize in providing objective reviews in certain fields. These reviewers should not themselves work in the field and should have no personal incentive to take sides. Try to reach agreements with other institutions on the number of such positions.

SUPPORT THE PUBLICATION OF CRITICISM AND NEGATIVE RESULTS. Criticism of other people's work or negative results are presently underappreciated. But these contributions are absolutely essential for the scientific method to work. Find ways to encourage the publication of such communication, for example by dedicated special issues.

OFFER COURSES ON SOCIAL AND COGNITIVE BIASES. This should be mandatory for everybody who works in academic research. We are part of communities and we have to learn about the associated pitfalls. Sit together with people from the social sciences, psychology, and the philosophy of science, and come up with proposals for lectures on the topic.

ALLOW A DIVISION OF LABOR BY SPECIALIZATION IN TASK. Nobody is good at everything, so don't expect scientists to be. Some are good reviewers, some are good mentors, some are good leaders, and some are skilled at science communication. Allow them to shine in what they're good at and make best use of it, but don't require the person who spends most evenings in student Q&As to also bring in loads of grant money. Offer people specific titles, degrees, or honors.

As a science writer or member of the public

ASK QUESTIONS. You're used to asking about conflicts of interest due to funding from industry. But you should also ask about conflicts of interest due to short-term grants or employment. Does the scientists' future funding depend on producing the results they just told you about?

Likewise, you should ask if the scientists' chance of continuing their research depends on their work being popular among their colleagues. Does their present position offer adequate protection from peer pressure?

And finally, just like you are used to scrutinizing statistics, you should also ask whether the scientists have taken steps to address their cognitive biases. Have they provided a balanced account of pros and cons or have they just advertised their own research?

You will find that for almost all research in the foundations of physics the answer to at least one of these questions is no. This means you can't trust these scientists' conclusions. Sad but true.

NOTES

Chapter 1: The Hidden Rules of Physics

1. Barbieri R, Giudice GF. 1988. "Upper bounds on supersymmetric particle masses." *Nucl. Phys. B*. 306:63.

2. Hooper D. 2008. *Nature's blueprint*. New York: Harper Collins.

3. Forshaw J. 2012. "Supersymmetry: is it really too good not to be true?" *Guardian*, December 9, 2012.

4. Lykken J, Spiropulu M. 2014. "Supersymmetry and the crisis in physics." *Scientific American*, May 1, 2014.

5. This is the Coleman-Mandula theorem. Coleman S, Mandula J. 1967. "All possible symmetries of the S matrix." *Phys. Rev.* 159:1251.

6. Haag R, Łopuszański J, Sohnius M. 1975. "All possible generators of supersymmetries of the S-matrix." *Nucl. Phys. B*. 88:257. The assumptions can be relaxed even further, but nothing physically interesting seems to have come out of it; see, e.g., Lykken J. 1996. *Introduction to supersymmetry*. FERMILAB-PUB-96/445-T. arXiv: hep-th/9612114.

7. For example, constraints on flavor-changing neutral currents and the electric dipole moment. See, e.g., Cohen AG, Kaplan DB, Nelson AE. 1996. "The more minimal supersymmetric standard model." *Phys. Lett. B*. 388: 588–598. arXiv:hep-ph/9607394.

8. Giudice GF. 2008. *Naturally speaking: the naturalness criterion and physics at the LHC*. arXiv:0801.2562 [hep-ph].

9. Arkani-Hamed N, Dimopoulos S, Dvali G. 1998. "The hierarchy problem and new dimensions at a millimeter." *Phys. Lett. B*. 429:263–272. arXiv:hep-ph/9803315.

10. Or even earlier to Nordström in 1905, though it's usually accredited to Kaluza and Klein since Nordström didn't work with general relativity, which was unknown at the time.

Chapter 2: What a Wonderful World

1. Chandrasekhar S. 1990. *Truth and beauty: aesthetics and motivations in science*. Chicago: University of Chicago Press; Orrell D. 2012. *Truth or beauty: science and the quest for order*. New Haven, CT: Yale University Press; Steward I. 2008. *Why beauty is truth: a history of symmetry*. New York: Basic Books; Kragh H. 2011. *Higher speculations*. Oxford, UK: Oxford University Press; McAllister JW. 1996. *Beauty and revolution in science*. Ithaca, NY: Cornell University Press.

2. Galileo G. 1967. *Dialogue concerning the two chief world systems, Ptolemaic and Copernican*. 2nd ed. Berkeley: University of California Press.

3. Quoted in McAllister JW. 1996. *Beauty and revolution in science*. Ithaca, NY: Cornell University Press, p. 178.

4. Newton I. 1729. *The general scholium*. Motte A, trans. https://isaac-newton.org/general-scholium.

5. Leibniz G. 1686. *Discourse on metaphysics*. Montgomery GR, trans. In: *Leibniz* (1902). La Salle, IL: Open Court.

6. Called "the principle of least action."

7. Quoted by Freeman Dyson in Weyl's obituary, *Nature*, March 10, 1956.

8. Rebsdorf S, Kragh H. 2002. "Edward Arthur Milne—the relations of mathematics to science." *Studies in History and Philosophy of Modern Physics* 33:51–64.

9. Quoted in Kragh H. 1990. *Dirac: a scientific biography*. Cambridge, UK: Cambridge University Press, p. 277.

10. Dalitz RH. 1987. "A biographical sketch of the life of Professor P. A. M. Dirac, OM, FRS." In: Taylor JG, Hilger A, editors. *Tributes to Paul Dirac*. Bristol, UK: Adam Hilger, p. 20. Quoted in McAllister JW. 1996. *Beauty and revolution in science*. Ithaca, NY: Cornell University Press, p. 16.

11. Kragh H. 1990. *Dirac: a scientific biography*. Cambridge, UK: Cambridge University Press, p. 292.

12. Einstein A. 2009. *Einstein's essays in science*. Harris A, trans. Mineola, NY: Dover Publications, pp. 17–18.

13. Poincaré H. 2001. *The value of science: the essential writings of Henri Poincaré*. Gould SJ, editor. New York: Modern Library, p. 369.

14. Poincaré H. 2001. *The value of science: the essential writings of Henri Poincaré*. Gould SJ, editor. New York: Modern Library, pp. 396–398.

15. Letter of Heisenberg to Einstein, in Heisenberg W. 1971. *Physics and beyond: encounters and conversations*. New York: HarperCollins, p. 68.

16. Heisenberg E. 1984. *Inner exile: recollections of a life with Werner Heisenberg*. Boston: Birkhäuser, p. 143.

17. Zee A. 1986. *Fearful symmetry: the search for beauty in modern physics*. New York: Macmillan.

18. Lederman L. 2006. *The God particle*. Boston: Mariner Books, p. 15.

19. The quark model was discovered independently at almost the same time by George Zweig.

20. Gell-Mann M, in a TED talk filmed March 2007. www.ted.com/talks/murray_gell_mann_on_beauty_and_truth_in_physics. The version I quoted is what is written on his slide. What he says is: "We have this remarkable experience in this field of fundamental physics that beauty is a very successful criterion to choose the right theory."

21. Lederman L. 2007. "The God particle et al." *Nature* 448:310–312.

22. Weinberg S. 2003. Interview with *Nova* (PBS), conducted by Joe McMaster. www.pbs.org/wgbh/nova/elegant/view-weinberg.html.

23. Wilczek F. 2015. *A beautiful question: finding nature's deep design*. New York: Penguin Press, p. 9

24. Interview conducted with author.

25. Greene B. 1999. *The elegant universe: superstrings, hidden dimensions, and the quest for the ultimate theory*. New York: WW Norton, p. 167.

26. Heisenberg E. 1984. *Inner exile: recollections of a life with Werner Heisenberg*. Boston: Birkhäuser, p. 144.

27. Schrödinger E. 1926. "Über das Verhältnis der Heisenberg-Bohr-Jordanschen Quantentheorie zu der meinen." *Ann Physik* 4(7):734–756.

28. Pauli W. 1979. "Wissenschaftlicher Briefwechsel mit Bohr, Einstein, Heisenberg u.a.: Band 1: 1919–1929." In: Hermann A, v. Meyenn K, Weisskopf VF, editors. *Sources in the History of Mathematics and Physical Sciences*. New York: Springer, p. 262.

29. Lemaître G. 1958. "Rencontres avec A. Einstein." *Revue des Questions Scientifiques* 129:129–132. Cited in Nussbaumer H. 2014. "Einstein's conversion from his static to an expanding universe." *EPJH* 39:37–62.

30. Obituary, Professor Sir Fred Hoyle. *Telegraph*, August 22, 2001.

31. Curtis A. 2012. "A mile or two off Yarmouth." *Adam Curtis: the medium and the message*, February 24, 2012. www.bbc.co.uk/blogs/adamcurtis/entries/512cde83-3afb-3048-9ece-dba774b10f89.

32. Vortex theory has been covered in: Kragh H. 2002. "The vortex atom: a Victorian theory of everything." *Centaurus* 44:32–114; Kragh H. 2011. *Higher speculations*. Oxford, UK: Oxford University Press.

33. Lodge O. 1883. "The ether and its functions." *Nature* 34:304–306, 328–330. Quoted in McAllister JW. 1996. *Beauty and revolution in science*. Ithaca, NY: Cornell University Press, p. 91.

34. Michelson AA. 1903. *Light waves and their uses*. Chicago: University of Chicago Press. Quoted in Kragh H. 2011. *Higher speculations*. Oxford, UK: Oxford University Press, p. 52.

35. Maxwell JC. 1875. "Atoms." In: *Encyclopaedia Britannica*. 9th ed. Quoted in Maxwell JC. 2011. *The scientific papers of James Clerk Maxwell*. Vol. 2. Niven WD, editor. Cambridge, UK: Cambridge University Press, p. 592.

36. Quoted in Kragh H. 2011. *Higher speculations*. Oxford, UK: Oxford University Press, p. 87.

37. Klein MJ. 1973. "Mechanical explanation at the end of the nineteenth century." *Centaurus* 17:72. Quoted in McAllister JW. 1996. *Beauty and revolution in science*. Ithaca, NY: Cornell University Press, p. 88.

38. Dirac P. 1951. "A new classical theory of electrons." *Proc R Soc Lond A.* 209:251.

39. Quoted in Kragh H. 1990. *Dirac: a scientific biography*. Cambridge, UK: Cambridge University Press, p. 184.

40. Only two physicists have won a Nobel Prize twice, and that mainly for experimental work: Maria Sklodowska-Curie in 1903 for the discovery of radioactivity and in 1911 for isolating pure radium (the latter prize was in the field of chemistry), and John Bardeen in 1956 for the invention of the transistor and again in 1972, together with Leon Cooper and John Schrieffer, for their jointly developed theory of superconductivity.

    Some of my colleagues argue that Bardeen should count as a theoretical physicist. Bardeen graduated in electrical engineering and both of his Nobel Prizes were for applied science, so I think that's quite a stretch. But since "theoretical physicist" is not a sharply defined term, I also think debating the issue is moot. Either way, it's irrelevant to my point. Bardeen's achievements clearly weren't based on the arguments from beauty and naturalness that this book is about.

41. Dawid R. 2013. *String theory and the scientific method*. Cambridge, UK: Cambridge University Press.

42. Kragh H. 2002. "The vortex atom: a Victorian theory of everything." *Centaurus* 44:32–114; Kragh H. 2011. *Higher speculations*. Oxford, UK: Oxford University Press.

43. Ellis G, Silk J. 2014. "Scientific method: defend the integrity of physics." *Nature* 516:321–323.

44. Dawid R. 2013. *String theory and the scientific method*. Cambridge, UK: Cambridge University Press, back flap.

45. Wagner CEM. 2005. "Lectures on supersymmetry (II)." PowerPoint presentation, Fermilab, Batavia, IL, June 23 and 30, 2005. www-cdf.fnal .gov/physics/lectures/Wagner_Lecture2.pdf.

46. Lederman L. 2007. "The God particle et al." *Nature* 448:310–312.

Chapter 3: The State of the Union

1. O'Brien FJ. 1858. *The diamond lens*. Project Gutenberg. www.gutenberg .org/ebooks/23169.

2. Strictly speaking, it's not the Higgs boson that gives masses to elementary particles but the non-vanishing background value of the Higgs field. The masses of composite particles like neutrons and protons are mostly binding energy and not due to the Higgs.

3. Glashow S. 1988. *Interactions: a journey through the mind of a particle physicist and the matter of this world*. New York: Warner Books.

4. An excellent overview on telescope technology and development can be found in Graham-Smith F. 2016. *Eyes on the sky: a spectrum of telescopes*. Oxford, UK: Oxford University Press.

5. The most amazing example may be the Hubble Ultra Deep Field. See Beckwith SVW et al. 2006. "The Hubble Ultra Deep Field." *Astron J.* 132: 1729–1755. arXiv:astro-ph/0607632. Or do a Google image search for "Hubble Ultra Deep Field."

6. Some cosmologists refer to the concordance model as the "standard model of cosmology." I will not use this nomenclature to prevent confusion with the standard model of particle physics, which is normally referred to as just the "standard model."

7. A temperature of 10^{17} Kelvin is pretty much the same as 10^{17} degrees Celsius and 10^{17} degrees Fahrenheit.

8. Lots could be said about the concordance model and its development, but I don't want to stray too far from the story. For further details, I recommend Gates E. 2010. *Einstein's telescope: the hunt for dark matter and dark energy in the universe*. New York: WW Norton; Siegel E. 2015. *Beyond the Galaxy: how humanity looked beyond our Milky Way and discovered the entire universe*. Hackensack, NJ: World Scientific Publishing.

9. Why do we call some things a model and other things a theory? There's no strict nomenclature for this; some names just stick, others don't. But loosely speaking, a theory is a general mathematical framework, whereas a model contains all the details you need to make a calculation. There are many different quantum field theories, for example, but the standard model specifies exactly which one. And the theory of general relativity alone doesn't tell you what to calculate; for this you also need to know what matter and energy populate space-time, like the concordance model does.

10. Portell X. 2011. "SUSY searches at the Tevatron and the LHC." Talk given at Physics in Collision, Vancouver, Canada, August/September 2011, slide 41. http://indico.cern.ch/event/117880/contributions/1330772/attachments/58548/84276/portell_SUSYsearches.pdf.

11. Allanach B. 2015. "Hint of new particle at CERN's Large Hadron Collider?" *Guardian*, December 16, 2015.

12. Quoted in Cho A. 2007. "Physicists' nightmare scenario: the Higgs and nothing else." *Science* 315:1657–1658.

13. Gross D. 2013. "Closing remarks." Presentation at Strings 2013, Sogang University, Seoul, South Korea, June 24–29, 2013. www.youtube.com/embed/vtXAwk1vkmk.

Chapter 4: Cracks in the Foundations

1. Kaku M. 1994. *Hyperspace: a scientific odyssey through parallel universes, time warps, and the tenth dimension*. Oxford, UK: Oxford University Press, p. 126; Hawking S. 2002. "Gödel and the end of the universe." Public lecture, Cambridge, UK, July 20, 2002; Strassler M. 2013. "Looking beyond the standard model." Lecture at the Higgs Symposium, University of Edinburgh, Edinburgh, Scotland, January 9–11, 2013. https://higgs.ph.ed.ac.uk/sites/default/files/Strassler_Looking%20Beyond%20SM.pdf; Greene B. 1999. *The elegant universe: superstrings, hidden dimensions, and the quest for the ultimate theory*. New York: WW Norton, p. 143. Davies P. 2007. *The Goldilocks enigma: why is the universe just right for life?* New York: Penguin, p. 101.

2. Koide Y. 1993. "Should the renewed tau mass value 1777 MeV be taken seriously?" *Mod Phys Lett*. A8:2071.

3. The mixing matrices actually collect the amplitudes, not the probabilities. There is one mixing matrix for neutrinos and one for down-type quarks. The latter is known as the Cabibbo-Kobayashi-Maskawa (CKM) matrix.

4. Peccei RD, Quinn HR. 1977. "CP conservation in the presence of pseudoparticles." *Phys Rev Lett*. 38(25):1440–1443.

5. Brahe T. 1602. *Astronomiae instauratae progymnasmata*. Quoted in Blair A. 1990. "Tycho Brahe's critique of Copernicus and the Copernican system." *Journal of the History of Ideas* 51(3):364.

6. For a good overview, see Barrow JD. 1981. "The lore of large numbers: some historical background to the anthropic principle." *Q Jl R Astr Soc*. 22:388–420.

7. This is a funky way of saying the coefficients should be close to 1.

8. Steigman G. 1978. "A crucial test of the Dirac cosmologies." *Ap J*. 221:407–411.

9. Though this wasn't understood until physicists learned how to deal with the harmless infinities of quantum field theories.

10. The definition of technical naturalness goes back to Gerard 't Hooft, who postulated that whenever there is a conspicuously small number in a quantum field theory, there must be a symmetry when the number is zero so that the smallness is protected by the symmetry, though this original definition has now been generalized to mean that there must be some explanation for small numbers—symmetry or something else. See: 't Hooft G et al. 1980. *Recent developments in gauge theories*. New York: Plenum Press, p. 135.

11. Glashow SL, Iliopoulos J, Maiani L. 1970. "Weak interactions with lepton–hadron symmetry." *Phys Rev.* 2(7):1285–1292.

12. Nadis S, Yau S-T. 2015. *From the Great Wall to the great collider: China and the quest to uncover the inner workings of the universe*. Somerville, MA: International Press of Boston, p. 75.

13. Randall L. 2006. *Warped passages: unraveling the mysteries of the universe's hidden dimensions*. New York: Harper Perennial, p. 253.

14. Interview conducted with author.

15. One of the supersymmetric particles, a superpartner of the standard-model particle that's called a gluon.

16. Thomas KD. 2015. "Beyond the Higgs: from the LHC to China." *The Institute Letter*, Summer 2015. Princeton, NJ: Institute for Advanced Study, p. 8.

Chapter 5: Ideal Theories

1. Voss RF, Clarke J. 1975. "'1/*f* noise' in music and speech." *Nature* 258:317–318.

2. Bach, of course, didn't start composing at the Big Bang, and humans can hear only a small range of frequencies. So that there are correlations over "all" time scales merely means during the duration of the composition and in the audible frequency range. Real music therefore doesn't actually have a perfect 1/*f* spectrum, but only approximates it in some range. Still, the universality is an interesting find.

3. Not to be confused with the area of philosophy that goes by the same name—the name is the only thing they have in common.

4. Einstein A. 1999. *Autobiographical notes*. La Salle, IL: Open Court, p. 63.

5. Weinberg S. 2015. *To explain the world: the discovery of modern science*. New York: Harper.

6. Wolchover N, Byrne P. 2014. "How to check if your universe exists." *Quanta Magazine*, July 11, 2014.

7. Davies P. 2007. "Universes galore: where will it all end?" In: Carr B, editor. *Universe or multiverse*. Cambridge, UK: Cambridge University Press, p. 495.

8. Ellis G. 2011. "Does the multiverse really exist?" *Scientific American*, August 2011, p. 40.

9. "Lawrence Krauss owned by David Gross on the multiverse religion." YouTube video, published June 26, 2014. www.youtube.com/watch?v=fEx5rWfz2ow.

10. Wells P. 2013. "Perimeter Institute and the crisis in modern physics." *MacLean's*, September 5, 2013.

11. Horgan J. 2011. "Is speculation in multiverses as immoral as speculation in subprime mortgages?" *Cross-check* (blog). *Scientific American*, January 28, 2011. https://blogs.scientificamerican.com/cross-check/is-speculation-in-multiverses-as-immoral-as-speculation-in-subprime-mortgages/.

12. Gefter A. 2005. "Is string theory in trouble?" *New Scientist*, December 14, 2005.

13. Carr B. 2008. "Defending the multiverse." *Astronomy & Geophysics* 49(2):2.36–2.37.

14. Naff CF. 2014. "Cosmic quest: an interview with physicist Max Tegmark." TheHumanist.com, May 8, 2014.

15. Siegfried T. 2013. "Belief in multiverse requires exceptional vision." ScienceNews.org, August 14, 2013.

16. Eternal inflation was first proposed in 1983 by Paul Steinhardt and Alexander Vilenkin. A different model was proposed a few years later by Alan Guth and Andrei Linde.

17. See, for example, Mukhanov V. 2015. "Inflation without selfreproduction." *Fortschr Phys.* 63(1). arXiv:1409.2335 [astro-ph.CO].

18. Bousso R, Susskind L. 2012. "The multiverse interpretation of quantum mechanics." *Phys Rev D.* 85:045007. arXiv:1105.3796 [hep-th].

19. Tegmark M. 2008. "The mathematical universe." *Found Phys.* 38:101–150. arXiv:0704.0646 [gr-qc].

20. Garriga J, Vilenkin A, Zhang J. 2016. "Black holes and the multiverse." *JCAP* 02:064. arXiv:1512.01819 [hep-th].

21. Weinberg S. 2005. "Living in the multiverse." Talk presented at Expectations of a Final Theory, Trinity College, Cambridge, UK, September 2005. arXiv:hep-th/0511037.

22. For the multiverse, this issue goes under the name "measure problem." See, for example, Vilenkin A. 2012. "Global structure of the multiverse and the measure problem." *AIP Conf Proc* 1514:7. arXiv:1301.0121 [hep-th].

23. Hoyle F. 1954. "On nuclear reactions occurring in very hot stars. I. The synthesis of elements from carbon to nickel." *Astrophys J Suppl Ser.* 1:121.

24. Kragh H. 2010. "When is a prediction anthropic? Fred Hoyle and the 7.65 MeV carbon resonance." Preprint. http://philsci-archive.pitt.edu/5332.

25. See, for example, Barrow JD, Tipler FJ. 1986. *The anthropic cosmological principle*. Oxford, UK: Oxford University Press; Davies P. 2007. *Cosmic jackpot: why our universe is just right for life*. Boston: Houghton Mifflin Harcourt.

26. Harnik RD, Kribs GD, Perez G. 2006. "A universe without weak interactions." *Phys Rev D.* 74:035006. arXiv:hep-ph/0604027.

27. Loeb A. 2014. "The habitable epoch of the early universe." *Int J Astrobiol.* 13(4):337–339. arXiv:1312.0613 [astro-ph.CO].

28. Adams FC, Grohs E. 2016. "Stellar helium burning in other universes: a solution to the triple alpha fine-tuning problem." arXiv:1608.04690 [astro-ph.CO].

29. I don't find this very surprising. That theories with dozens of nonlinearly interacting components give rise to complex structure is something I expect to be the rule rather than the exception.

30. Don Page argues against it here: Page DN. 2011. "Preliminary inconclusive hint of evidence against optimal fine tuning of the cosmological constant for maximizing the fraction of baryons becoming life." arXiv:1101.2444.

31. Martel H, Shapiro PR, Weinberg S. 1998. "Likely values of the cosmological constant." *Astrophys J.* 492:29. arXiv:astro-ph/9701099.

32. Rumors that the cosmological constant was not zero and instead positive-valued had been around as early as a decade previously, based on other measurements that, however, remained somewhat inconclusive. See, for example, Efstathiou G, Sutherland WJ, Maddox SJ. 1990. "The cosmological constant and cold dark matter." *Nature* 348:705–707; Krauss LM, Turner MS. 1995. "The cosmological constant is back." *Gen Rel Grav.* 27:1137–1144. arXiv:astro-ph/9504003. Weinberg had been thinking about an anthropic explanation for the cosmological constant even earlier; see Weinberg S. 1989. "The cosmological constant problem." *Rev Mod Phys.* 61(1):1–23.

33. Johnson G. 2014. "Dissonance: Schoenberg." *New York Times*, video, May 30, 2014. www.nytimes.com/video/arts/music/100000002837182/dissonance-schoenberg.html.

34. McDermott JH, Schultz AF, Undurraga EA, Godoy RA. 2016. "Indifference to dissonance in native Amazonians reveals cultural variation in music perception." *Nature* 535:547–550.

Chapter 6: The Incomprehensible Comprehensibility of Quantum Mechanics

1. This section's title paraphrases the comedian Louis C.K., who said, "Everything is amazing right now, and nobody's happy." See "Louis CK everything is amazing and nobody is happy." YouTube video, published October 24, 2015. www.youtube.com/watch?v=q8LaT5Iiwo4.

2. Popescu S. 2014. "Nonlocality beyond quantum mechanics." *Nature Physics* 10:264–270.

3. Quoted in Wolchover N. 2014. "Have we been interpreting quantum mechanics wrong this whole time?" *Wired*, June 30, 2014.

4. Weinberg S. 2012. *Lectures on quantum mechanics*. Cambridge, UK: Cambridge University Press.

5. Scheidl T et al. 2010. "Violation of local realism with freedom of choice." *Proc Natl Acad Sci USA*. 107:19708. arXiv:0811.3129 [quant-ph].

6. An excellent book for further reading is Musser G. 2015. *Spooky action at a distance: the phenomenon that reimagines space and time—and what it means for black holes, the big bang, and theories of everything*. New York: Farrar, Straus and Giroux.

7. Mermin ND. 1989. "What's wrong with this pillow?" *Physics Today*, April 1989. And no, it wasn't Feynman who said it; see Mermin ND. 2004. "Could Feynman have said this?" *Physics Today*, May 2004.

8. Vaidman L. 2012. "Time symmetry and the many worlds interpretation." In: Saunders S, Barrett J, Kent A, Wallace D, editors. *Many worlds? Everett, quantum theory, and reality*. Oxford, UK: Oxford University Press, p. 582; Tegmark M. 2003. "Parallel universes." *Scientific American*, May 2003.

9. Mermin ND. 2016. *Why quark rhymes with pork and other scientific diversions*. Cambridge, UK: Cambridge University Press; Carroll S. 2014. "Why the many-worlds formulation of quantum mechanics is probably correct." *Preposterous Universe*, June 30, 2014. http://www

.preposterousuniverse.com/blog/2014/06/30/why-the-many-worlds-formulation-of-quantum-mechanics-is-probably-correct/.

10. The different interpretations also suggest that the conditions to get the same results as the Copenhagen interpretation might not always be fulfilled, in which case both could be distinguished by experiment. This, however, hasn't happened so far.

11. Bell M, Gottfried K, Veltman M. 2001. *John S. Bell on the foundations of quantum mechanics*. River Edge, NJ: World Scientific Publishing, p. 199.

12. Polkinghorne J. 2002. *Quantum theory: a very short introduction*. Oxford, UK: Oxford University Press, p. 89.

13. Tegmark M. 2015. *Our mathematical universe: my quest for the ultimate nature of reality*. New York: Vintage, p. 187; Tegmark M. 2003. "Parallel universes." *Scientific American*, May 2003.

14. Kent A. 2014. "Our quantum problem." *Aeon*, January 28, 2014.

15. He mentioned he was working on a paper about this, which appeared two months later. See Weinberg S. 2016. "What happens in a measurement?" *Phys Rev A*. 93:032124. arXiv:1603.06008 [quant-ph].

16. That a state remains "pure" means quantum properties don't decohere.

17. Orzel C. 2010. *How to teach quantum physics to your dog*. New York: Scribner.

18. William Daniel Phillips, who won the 1997 Nobel Prize in Physics together with Claude Cohen-Tannoudji and Steven Chu for laser cooling, a technique to slow down atoms.

19. Sparkes A et al. 2010. "Towards robot scientists for autonomous scientific discovery." *Automated Experimentation* 2:1.

20. Schmidt M, Lipson H. 2009. "Distilling free-form natural laws from experimental data." *Science* 324:81–85.

21. Krenn M, Malik M, Fickler R, Lapkiewicz R, Zeilinger A. 2016. "Automated search for new quantum experiments." *Phys Rev Lett.* 116:090405.

22. Quoted in Ball P. 2016. "Focus: computer chooses quantum experiments." *Physics* 9:25.

23. Powell E. 2011. "Discover interview: Anton Zeilinger dangled from windows, teleported photons, and taught the Dalai Lama." *Discover Magazine*, July–August 2011.

24. You can download the game here: www.scienceathome.org /games/quantum-moves/game.

25. Sørensen JJWH et al. 2016. "Exploring the quantum speed limit with computer games." *Nature* 532:210–213.

26. It goes back to this quote: "We often discussed his notions on objective reality. I recall that during one walk Einstein suddenly stopped, turned to me and asked whether I really believed that the moon exists only when I look at it." In: Pais A. 1979. "Einstein and the quantum theory." *Rev Mod Phys.* 51:907.

27. Wigner EP. 1960. "The unreasonable effectiveness of mathematics in the natural sciences." *Comm Pure Appl Math.* 13:1–14.

Chapter 7: One to Rule Them All

1. See Chapter 2.

2. It is not the mixing angle itself that is 3/8 but its sine square.

3. Hooper D. 2008. *Nature's blueprint*. New York: Harper Collins, p. 193.

4. Castelvecchi D. 2012. "Is supersymmetry dead?" *Scientific American*, May 2012.

5. Gross D. 2005. "Einstein and the search for unification." *Current Science* 89(12):25.

6. Quoted in Cho A. 2007. "Physicists' nightmare scenario: the Higgs and nothing else." *Science* 315:1657–1658.

7. Wilczek F. 2016. "Power over nature." *Edge*, April 20, 2016. https://www.edge.org/conversation/frank_wilczek-power-over-nature.

8. McAllister JW. 1996. *Beauty and revolution in science*. Ithaca, NY: Cornell University Press.

9. Mulvey PJ, Nicholson S. 2014. "Trends in physics PhDs." *Focus On*, February 2014. College Park, MD: AIP Statistical Research Center.

10. Forman P, Heilbron JL, Weart S. 1975. "Physics circa 1900." In: McCormmach R, editor. *Historical Studies in the Physical Sciences, Volume 5*. Princeton, NJ: Princeton University Press.

11. Sinatra R et al. 2015. "A century of physics." *Nature Physics* 11:791–796.

12. Larsen PO, von Ins M. 2010. "The rate of growth in scientific publication and the decline in coverage provided by Science Citation Index." *Scientometrics* 84(3):575–603.

13. Sinatra R et al. 2015. "A century of physics." *Nature Physics* 11:791–796.

14. Palmer CL, Cragin MH, Hogan TP. 2004. "Information at the intersections of discovery: Case studies in neuroscience." *Proceedings of the Association for Information Science and Technology* 41:448–455.

15. Uzzi B, Mukherjee S, Stringer M, Jones B. 2013. "Atypical combinations and scientific impact." *Science* 342(6157):468–472.

16. Sinatra R et al. 2015. "A century of physics." *Nature Physics* 11:791–796.

17. King C. 2016. "Single-author papers: a waning share of output, but still providing the tools for progress." *Science Watch*, retrieved January 2016. http://sciencewatch.com/articles/single-author-papers-waning-share-output-still-providing-tools-progress.

18. Maher B, Anfres MS. 2016. "Young scientists under pressure: what the data show." *Nature* 538:44–45.

19. von Hippel T, von Hippel C. 2015. "To apply or not to apply: a survey analysis of grant writing costs and benefits." *PLoS One* 10(3):e0118494.

20. Chubba J, Watermeyer R. 2016 Feb. "Artifice or integrity in the marketization of research impact? Investigating the moral economy of (pathways to) impact statements within research funding proposals in the UK and Australia." *Studies in Higher Education* 42(12):2360–2372.

21. The essence of tenure is a contract that has no end date and can't be terminated without very good reasons. Very good reasons are, for example, misconduct or misbehavior, but not, say, holding an annoying scientific opinion. It works similarly in most countries. The purpose of tenure is (or should I say "was"?) to protect academics from being dismissed for pursuing unpopular or provocative research topics. I sometimes stumble upon people (mostly Americans) who seem to think it's an unfair privilege if academics enjoy such job security. But that's a pointless criticism because without this security academic research runs a high risk of wasting tax money.

22. Shulman S, Hopkins B, Kelchen R, Mastracci S, Yaya M, Barnshaw J, Dunietz S. 2016. "Higher education at a crossroads: the economic value of tenure and the security of the profession." *Academe*, March–April 2016, pp. 9–23. https://www.aaup.org/sites/default/files/2015-16EconomicStatusReport.pdf.

23. Brembs B. 2015. "Booming university administrations." *Björn.Brembs.Blog*, January 7, 2015. http://bjoern.brembs.net/2015/01/booming-university-administrations.

24. Foster JG et al. 2015. "Tradition and innovation in scientists' research strategies." *Am Sociol Rev.* 80(5):875–908.

25. Weaver K, Garcia SM, Schwarz N, Miller DT. 2007. "Inferring the popularity of an opinion from its familiarity: a repetitive voice can sound like a chorus." *J Pers Soc Psychol.* 92(5):821–833.

26. Odenwald S. 2015. "The future of physics." *Huffington Post*, January 26, 2015. https://www.huffingtonpost.com/dr-sten-odenwald/the-future-of-physics_b_6506304.html.

27. Lykken J, Spiropulu M. 2014. "Supersymmetry and the crisis in physics." *Scientific American*, May 1, 2014.

28. Khosrovshahi GB. n.d. Interview with Alain Connes. Retrieved January 2016. www.freewebs.com/cvdegosson/connes-interview.pdf.

29. Kac M. 1966. "Can one hear the shape of a drum?" *Amer Math Monthly* 73(4):1–23.

30. Chamseddine AH, Connes A, Marcolli M. 2007. "Gravity and the standard model with neutrino mixing." *Adv Theor Math Phys.* 11:991–1089. arXiv:hep-th/0610241.

31. Connes A. 2008. "Irony." *Noncommutative Geometry*, August 4, 2008. http://noncommutativegeometry.blogspot.de/2008/08/irony .html.

32. Chamseddine AH, Connes A. 2012. "Resilience of the spectral standard model." *J High Energy Phys.* 1209:104. arXiv:1208.1030 [hep-ph].

33. Hebecker A, Wetterich W. 2003. "Spinor gravity." *Phys Lett B.* 574:269–275. arXiv:hep-th/0307109; Finster F, Kleiner J. 2015. "Causal fermion systems as a candidate for a unified physical theory." *J Phys. Conf Ser* 626:012020. arXiv:1502.03587 [math-ph].

34. The 1930s theory with additional dimensions of space that we briefly met in Chapter 1.

35. Lisi AG. 2007. "An exceptionally simple theory of everything." arXiv:0711.0770 [hep-th].

36. Lisi AG, Weatherall JO. 2010. "A geometric theory of everything." *Scientific American*, December 2010.

Chapter 8: Space, the Final Frontier

1. Absence of ghosts.

2. Kachru S, Kallosh R, Linde A, Trivedi SP. 2003. "De Sitter vacua in string theory." *Phys Rev.* D68:046005. arXiv:hep-th/0301240.

3. Hossenfelder S. 2013. "Whatever happened to AdS/CFT and the quark gluon plasma?" *Backreaction*, September 12, 2013. http://backreaction.blogspot.com/2013/09/whatever-happened-to-adscft-and-quark.html.

4. Conlon J. 2015. *Why string theory?* Boca Raton, FL: CRC Press, p. 135.

5. Dyson F. 2009. "Birds and frogs." *Notices of the AMS* 56(2):221.

6. A wonderful book about the string-math connection is Yau S-T, Nadis S. 2012. *The shape of inner space: string theory and the geometry of the universe's hidden dimensions*. New York: Basic Books.

7. An almost digestible account of this relation can be found in Ronan M. 2006. *Symmetry and the monster: the story of one of the greatest quests of mathematics*. Oxford, UK: Oxford University Press.

8. This relation is often more specifically referred to as the AdS/CFT duality to emphasize that the gravitational theory is in anti–de Sitter (AdS) space (a space-time with a negative cosmological constant) and the gauge theory is a (supersymmetric) conformal field theory (CFT) in a space with one dimension less, similar but not identical to the theories of the standard model.

9. Polchinski J. 2015. "String theory to the rescue." arXiv:1512.02477 [hep-th].

10. The cross-section for scattering W-bosons off each other.

11. Polchinski J. 2004. "Monopoles, duality, and string theory." *Int J Mod Phys*. A19S1:145–156. arXiv:hep-th/0304042.

12. The reference to "information" is somewhat of a misnomer because the problem does not depend on exactly what is meant by information. What causes the trouble is that black hole evaporation is fundamentally irreversible.

13. Almheiri A, Marolf D, Polchinski J, Sully J. 2013. "Black holes: complementarity or firewalls?" *J High Energy Phys*. 2013(2):62. arXiv:

1207.3123 [hep-th]. Or claimed to show—their proof relies on an unnecessary hidden assumption that can be dropped, making the whole problem vanish. See Hossenfelder S. 2015. "Disentangling the black hole vacuum." *Phys Rev D.* 91:044015. arXiv:1401.0288.

14. The other three are death, loneliness, and freedom. See Yalom ID. 1980. *Existential psychotherapy*. New York: Basic Books.

15. Greene A. 2014. "Stephen King: the Rolling Stone interview." *Rolling Stone*, October 31, 2014.

16. Conlon J. 2015. *Why string theory?* Boca Raton, FL: CRC Press, p. 236.

17. Witten E. 2003. "Viewpoints on string theory." *Nova: The Elegant Universe*. PBS. http://www.pbs.org/wgbh/nova/elegant/viewpoints.html.

18. Brink L, Henneaux M. 1988. *Principles of string theory*. Boston: Springer.

19. Greene B. 1999. *The elegant universe: superstrings, hidden dimensions, and the quest for the ultimate theory*. New York: WW Norton, p. 82.

20. The most unpopular one is my claim that we're trying to solve the wrong problem—what we need to understand is not what happens to gravity at short distances but what happens to quantization at short distances. See Hossenfelder S. 2013. "A possibility to solve the problems with quantizing gravity." *Phys Lett B.* 725:473–476. arXiv:1208.5874 [gr-qc].

21. Cheng Z, Wen X-G. 2012. "Emergence of helicity +/- 2 modes (gravitons) from qubit models." *Nuclear Physics B.* 863. arXiv:0907.1203 [gr-qc].

22. This isn't the only math problem with the standard model or quantum field theories more generally. Another such problem is Haag's theorem, which states that all quantum field theories are trivial and physically irrelevant. That's somewhat disturbing, so physicists ignore the theorem.

Chapter 9: The Universe, All There Is, and the Rest

1. Vinkers CH, Tijdink JK, Otte WM. 2015. "Use of positive and negative words in scientific PubMed abstracts between 1974 and 2014: retrospective analysis." *BMJ* 351:h6467.

2. In the United States, the income inequality in academia is now larger than in industry or government. See Lok C. 2016. "Science's 1%: how income inequality is getting worse in research." *Nature* 537:471–473.

3. This, importantly, means that bad scientific practices can become dominant even though no individual scientist changes his or her behavior. See, for example, Smaldino PE, McElreath R. 2016. "The natural selection of bad science." *Royal Society Open Science* 3:160384. arXiv:1605.09511 [physics.soc-ph].

4. Hoyle F. 1967. "Concluding remarks." *Proc Royal Soc A.* 301:171. Hoyle also relates that Baade used the opportunity to bet Pauli that the neutrino would be discovered. Baade finally got his due, in champagne, when Cowen and Reines reported the successful detection.

5. Freese K. 2014. *Cosmic cocktail: three parts dark matter*. Princeton, NJ: Princeton University Press.

6. Feng JL. 2008. "Collider physics and cosmology." *Class Quant Grav.* 25:114003. arXiv:0801.1334 [gr-qc]; Buckley RB, Randall L. 2011. "Xogenesis." *J High Energy Phys.* 1109:009. arXiv:1009.0270 [hep-ph].

7. Baudis L. 2015. "Dark matter searches." *Ann Phys (Berlin)* 528(1–2):74–83. arXiv:1509.00869 [astro-ph.CO].

8. Goodman MW, Witten E. 1985. "Detectability of certain dark-matter candidates." *Phys Rev D.* 31(12):3059–3063.

9. Gelmini G. 1987. "Bounds on galactic cold dark matter particle candidates and solar axions from a Ge-spectrometer." In: Hinchliffe I, editor. *Proceedings of the theoretical workshop on cosmology and particle physics: July 28–Aug. 15, 1986, Lawrence Berkeley Laboratory, Berkeley, California*. Singapore: World Scientific.

10. The DAMA collaboration has detected statistically significant events of unknown origin whose frequency periodically changes over the course of the year (Bernabei R. 2008. "First results from DAMA/LIBRA and the combined results with DAMA/NaI." *Eur Phys J.* C56:333–355. arXiv:0804.2741 [astro-ph]). They've been seeing this signal for more than a decade. Such an annual modulation is what we would expect from dark matter because the probability that a dark matter signal is detected depends on the direction from which dark matter particles fall in. Since Earth moves through the presumably existing dark matter cloud on its way around the Sun, the direction of the incoming dark matter flux changes over the course of the year. Unfortunately, other experiments have excluded that the DAMA signal can be caused by dark matter because if that were so, it should have shown up also in other detectors, which hasn't happened. Presently nobody knows what DAMA is detecting.

11. Benoit A et al. 2001. "First results of the EDELWEISS WIMP search using a 320 g heat-and-ionization Ge detector." *Phys Lett B.* 513:15–22. arXiv: astro-ph/0106094.

12. Xenon100 Collaboration. 2013. "Limits on spin-dependent WIMP-nucleon cross sections from 225 live days of XENON100 data." *Phys Rev Lett.* 111:021301. arXiv:1301.6620 [astro-ph.CO].

13. Some people in the field disagree that the parameter space of the WIMP miracle has been excluded and argue that parts of the parameter space are still allowed. This disagreement comes about because there isn't just one model for WIMPs but many different ones. The original WIMP had an interaction mediated by a weak boson and was classified by a specific symmetry group. This variant has been ruled out, but other ways to mediate the interaction or other types of particles are still compatible with current data.

14. That is so if axions are produced before inflation. If axions are produced after inflation, the density of the axion condensate could be just about right to make up dark matter. The condensate, however, is a patchwork of regions separated by boundaries called "domain walls," and the domain walls have an energy density much too high to be compatible with observations.

15. Milgrom M. 1983. "A modification of the Newtonian dynamics—implications for galaxies." *Astrophys J.* 270:371.

16. See, for example, Moffat JW, Rahvar S. 2013. "The MOG weak field approximation and observational test of galaxy rotation curves." *Mon Notices Royal Astron Soc.* 436:1439. arXiv:1306.6383 [astroph.GA]. See also other papers by these authors.

17. Berezhiani L, Khoury J. 2015. "Theory of dark matter superfluidity." *Phys Rev D.* 92:103510. arXiv:1507.01019 [astro-ph.CO]. A similar idea was previously proposed in Bruneton J-P, Liberati S, Sindoni L, Famaey B. 2009. "Reconciling MOND and dark matter?" *JCAP* 3:21. arXiv:0811.3143 [astro-ph].

18. If you look up the cosmological constant, the present value is quoted as Λ about $10^{-52}/\mathrm{m}^2$. This is not what particle physicists refer to as the cosmological constant scale. Instead, they use the energy density related to the cosmological constant, which is $c^4\Lambda/G$ (where G is Newton's constant and c is the speed of light), multiply it by $\hbar^3$, and take the fourth root of this, which results in about $10^4/\mathrm{m}$ or, by taking the inverse, in a distance scale of about 1/10 of a mm. In terms of distance scales, the discrepancy with the Planck length is about 30 orders of magnitude (see Figure 14). In terms of energy density, one has the fourth power of this, resulting in the—more frequently quoted but somewhat misleading—discrepancy of 120 orders of magnitude.

19. Boyle LA, Steinhardt PJ, Turok N. 2006. "Inflationary predictions for scalar and tensor fluctuations reconsidered." *Phys Rev Lett.* 96:111301. arXiv:astro-ph/0507455.

20. Gaussianity and TE correlations, to mention the most important ones.

21. Martin J, Ringeval C, Trotta R, Vennin V. 2014. "The best inflationary models after Planck." *JCAP* 1403:039. arXiv:1312.3529 [astro-ph.CO].

22. Azhar F, Butterfield J. 2017. "Scientific realism and primordial cosmology." In: Saatsi J, editor. *The Routledge handbook of scientific realism*. New York: Routledge. arXiv:1606.04071 [physics.hist-ph].

23. Silk J. 2007. "The dark side of the universe." *Astron Geophys.* 48(2):2.30–2.38.

24. Ijjas A, Steinhardt PJ, Loeb A. 2017. "Pop goes the universe." *Scientific American*, January 2017.

25. Hawking SW, Ellis GFR. 1973. *The large scale structure of space-time.* Cambridge, UK: Cambridge University Press.

26. Ellis GFR. 1975. "Cosmology and verifiability." *QJRAS* 16:245–264.

27. Stepnes S. 2007. "Detector challenges at the LHC." *Nature* 448:290–296.

28. Ellis GFR, Brundrit GB. 1979. "Life in the infinite universe." *QJRAS* 20:37–41.

29. Hilbert D. 1926. "Über das Unendliche." *Mathematische Annalen* 95. Berlin: Springer.

30. Krauss L. 2012. "The consolation of philosophy." *Scientific American*, April 27, 2012.

31. Weinberg S. 1994. *Dreams of a final theory: the scientist's search for the ultimate laws of nature*. New York: Vintage.

32. Hawking S, Mlodinow L. 2010. *The grand design*. New York: Bantam.

33. Pigliucci M. 2012. "Lawrence Krauss: another physicist with an anti-philosophy complex." *Rationally Speaking*, April 25, 2012. http://rationallyspeaking.blogspot.de/2012/04/lawrence-krauss-another-physicist-with.html.

34. Maudlin T. 2015. "Why physics needs philosophy." *The Nature of Reality* (blog), April 23, 2015. PBS. www.pbs.org/wgbh/nova/blogs/physics/2015/04/physics-needs-philosophy/.

Chapter 10: Knowledge Is Power

1. Hooper D. 2008. *Nature's blueprint*. New York: Harper Collins, p. 5.

2. Rutjens BT, Heine SJ. 2016. "The immoral landscape? Scientists are associated with violations of morality." *PLoS ONE* 11(4):e0152798.

3. Watson JD. 2001. *The double helix: a personal account of the discovery of the structure of DNA*. New York: Touchstone, p. 210.

4. Fleming PA, Bateman PW. 2016. "The good, the bad, and the ugly: which Australian terrestrial mammal species attract most research?" *Mammal Rev.* 46(4):241–254.

5. "I would venture that the main reason why these statistical time-series techniques have not been incorporated for use in official climate forecasts is related to aesthetics—a topic which I doubt comes up much at meetings of the IPCC." Orrell D. 2012. *Truth or beauty: science and the quest for order*. New Haven, CT: Yale University Press, p. 214.

6. Krugman P. 2009. "How did economists get it so wrong?" *New York Times*, September 2, 2009.

7. See, for example, Farmer JD, Geanakoplos J. 2008. "The virtues and vices of equilibrium and the future of financial economics." arXiv:0803.2996 [q-fin.GN].

8. Currently 3.8 σ. See Yue AT et al. 2013. "Improved determination of the neutron lifetime." *Phys Rev Lett.* 111:222501. arXiv:1309.2623 [nucl-ex].

9. Patrignani C et al. (Particle Data Group). 2016. "Review of particle physics." *Chin Phys C* 40:100001.

10. Baker M. 2016. "1,500 scientists lift the lid on reproducibility." *Nature* 533:452–454.

11. I think the firewall paradox is simply based on a faulty proof. See Hossenfelder S. 2015. "Disentangling the black hole vacuum." *Phys Rev D.* 91:044015. arXiv:1401.0288. Regardless of its status however, it's interesting to see what consequences my colleagues have drawn.

12. Maldacena J, Susskind L. 2013. "Cool horizons for entangled black holes." *Fortschr Physik* 61(9):781–811. arXiv:1306.0533 [hep-th].

13. The wormhole is known as the Einstein-Rosen (ER) bridge, and the entangled particles are also known as EPR states because they were first

discussed by Einstein, Podolsky, and Rosen (Einstein A, Podolsky B, Rosen N. 1935. "Can quantum-mechanical description of physical reality be considered complete?" *Phys Rev.* 47:777). The conjecture that the two describe the same connection has therefore entered the literature as ER = EPR.

14. Further reading: Banaji MR, Greenwald AG. 2013. *Blindspot: hidden biases of good people*. New York: Delacorte; Kahneman D. 2012. *Thinking, fast and slow*. New York: Penguin.

15. Balcetis E. 2008. "Where the motivation resides and self-deception hides: how motivated cognition accomplishes self-deception." *Social and Personality Psychology Compass* 2(1):361–381.

16. Okay, this isn't actually what he said, but it's a widely spread paraphrased version of "A new scientific truth does not triumph by convincing its opponents and making them see the light, but rather because its opponents eventually die, and a new generation grows up that is familiar with it." Planck M. 1948. *Wissenschaftliche Selbstbiographie. Mit einem Bildnis und der von Max von Laue gehaltenen Traueransprache*. Leipzig: Johann Ambrosius Barth Verlag, p. 22.

17. Stanovich KE, West RF. 2008. "On the relative independence of thinking biases and cognitive ability." *J Pers Soc Psychol.* 94(4):672–695.

18. Cruz MG, Boster FJ, Rodriguez JI. 1997. "The impact of group size and proportion of shared information on the exchange and integration of information in groups." *Communic Res.* 24(3):291–313.

19. Wolchover N. 2016. "Supersymmetry bet settled with cognac." *Quanta Magazine*, August 22, 2016.

Appendix A: The Standard Model Particles

1. There are many excellent books about the standard model and particle accelerators that go into more detail than is necessary for my purposes. Just to mention two recent ones: Carroll S. 2013. *The particle at the end*

of the universe: how the hunt for the Higgs boson leads us to the edge of a new world. New York: Dutton; Moffat J. 2014. *Cracking the particle code of the universe: the hunt for the Higgs boson*. Oxford, UK: Oxford University Press.

2. Eberhart O et al. 2012. "Impact of a Higgs boson at a mass of 126 GeV on the standard model with three and four fermion generations." *Phys Rev Lett.* 109:241802. arXiv:1209.1101 [hep-ph].

Appendix B: The Trouble with Naturalness

1. Part of physicists' reliance on the decoupling of scales goes back to a 1975 paper that proves decoupling in quantum field theories under certain circumstances. The proof, however, relies on renormalizability and uses a mass-dependent renormalization scheme, both of which are questionable assumptions. See Appelquist T, Carazzone J. 1975. "Infrared singularities and massive fields." *Phys Rev.* D11:28565.

2. Anderson G, Castano D. 1995. "Measures of fine tuning." *Phys Lett B.* 347:300–308. arXiv:hep-ph/9409419.

3. For more details, see Hossenfelder S. 2018. "Screams for explanation: finetuning and naturalness in the foundations of physics." arXiv:1801.02176.

INDEX

Page numbers in **bold** indicate definition/explanation of a term.

ABOUT THE AUTHOR

Joerg Steinmetz

SABINE HOSSENFELDER is a research fellow at the Frankfurt Institute for Advanced Studies and the author of the popular physics blog *Backreaction*. She has written for *New Scientist*, *Scientific American*, and *NOVA*. She lives in Heidelberg, Germany.